Kites can be deceptively simple, surprisingly complex, and lots of fun

## About the Author:

Glenn Davison designs, builds, and flies all kinds of kites.

He has written several kite books, "Kites in the Classroom," "How to Fly a Kite," "Miniature Kites," "Indoor Kite Flying," the "Kite Workshop Handbook," and "Building Free and Recycled Kites."

He is the president of a kite club, past Director of the American Kitefliers Association, past Chairman of the Kite Education Committee, a member of the Tethered Aerosystems Working Group, and has participated in hundreds of kite events.

## Dedication:

This book is dedicated to all of us who have looked up in wonder at everything that flies including birds, planes, and kites of all shapes, sizes, styles, colors, and nationalities.

Thanks to Geoff Bland at NASA for encouraging me to write this book and for his contributions.

## Overview:

This book describes the physics of kites and offers many related experiments.

## Safety:

You are responsible for your own safety. Never fly a kite near cars, roads, power lines, telephone poles, airplanes, trees, or people. Never fly under bad conditions. Be safe at all times.

(c) 2020 Glenn Davison, all rights reserved.

# Contents

Kite myths and mistakes .................................................................. 5
Kite styles and components ............................................................. 8
    Sail ............................................................................................... 11
    Framework (with spars) .............................................................. 14
    Bridles ......................................................................................... 17
    Tails ............................................................................................. 20
    Line (or string) ............................................................................ 27
    Build a simple kite model ........................................................... 33
    Build a kite flying machine ......................................................... 37
The force of wind ........................................................................... 39
    What causes wind? .................................................................... 40
    Wind speed ................................................................................ 41
    Wind direction ........................................................................... 46
    Wind turbulence ........................................................................ 48
    Wind shadow ............................................................................. 50
Forces on a kite .............................................................................. 52
    Concept: Thrust ......................................................................... 53
    Concept: Drag ............................................................................ 54
    Concept: Roll, pitch, and yaw .................................................... 59
    Concept: Lift .............................................................................. 60
    Concept: Gravity ........................................................................ 69
    Concept: Stiffness vs. flexibility ................................................. 73
    Concept: Airflow over the kite ................................................... 77
    Concept: Angle of attack ........................................................... 78

- Concept: Flying angle ........................................................ 79
- Concept: Tension ............................................................... 80
- Concept: Gliding vs. stalling ............................................ 81
- Concept: Inverse square law............................................ 83
- How do I make my kite fly better?..................................... 84
  - Stability is important...................................................... 85
  - Methods to increase kite stability................................. 88
  - Adjust the bow or dihedral angle.................................. 91
  - Adjust for stability vs. lift .............................................. 93
  - Check the symmetry ...................................................... 94
  - Check the balance .......................................................... 96
  - Case studies and their solutions .................................. 97
  - Actions taken in the field ............................................ 100
- Uses and applications of kites........................................... 101
  - Measuring the atmosphere ......................................... 101
  - Parachute releases........................................................ 102
  - Kite Aerial Photography ............................................... 104
  - Wind instruments ......................................................... 105
  - Line laundry................................................................... 106
  - Line climbers ................................................................. 107
  - Ham Radio Antenna ..................................................... 108
  - Kites and balloons ........................................................ 109
- Summary of kite physics ..................................................... 110
- Glossary of terms ................................................................. 112
- Kite physics resources ......................................................... 113
- Bibliography .......................................................................... 114

## Kite myths and mistakes

People have wild ideas about kites and what makes them fly. People frequently mistake many things for kites such as: tails, windsocks, round parachutes, and various decorations that dangle from a kite line. Anything you see in the air that is attached to a string is not guaranteed to be a kite.

There are many wild theories about the contributions of momentum, rising air, the earth's rotation, gliding, falling, torque, and leverage. Often, these guesses have no basis in physics or facts.

One myth is, "If you bridle it right, anything will fly." That's cute, but it's not true.

Another myth, "If it flies, the Bernoulli Principle explains it." That's not true but it is widely believed to be true.

Kites are not magic. They are part of a system that results in flight. The system includes wind, a tether, and the kite itself. The forces are the same for all kites. The wind lifts the kite, while the tether prevents it from blowing away.

Let's investigate!

**Kite Definition**

> A tethered object with an aerodynamic surface that generates lift in order to overcome gravity and fly.

Kites have a long history of contributing to science, education, and flight. A kite is the perfect vehicle for Science, Technology, Engineering, and Math (STEM) because:

- Kites provide a transformative hook to learning through:
    - art, craft, design skills
    - engineering, mathematics
    - science, physics, aeronautics, aerodynamics
    - history, language, poetry, and writing
- Kites are an enjoyable way to inject STEM into other events such as community activities, cultural activities, and school events.
- Kites are a fun activity that can be added to the learning experience with great results.
- Flying kites is an activity that can be enjoyed by people of all ages throughout the year.

Today, there are many activities that involve kites. We fly kites:

- Outdoors all year long
- Indoors with no wind (Really!)
- At night with lights
- To take photos, video
- By people of all ages
- In nearly every country worldwide

But how do kites fly and what makes them fly well? That will be our focus.

## Kite styles and components

There are a wide variety of kite styles that originate in countries all over the world. Japan alone has over 700 styles of kites. Kite styles include: flat, bowed, flexible, cellular box kites, dihedral, airfoil (air-inflated), rotary, single-line, dual-line, quad-line, miniature, giant, and traditional. There are power kites that pull people on land, sea, ice, and snow.

These can be tall, square, wide, narrow, and three dimensional in many combinations.

A surface is required and a tether or flying line is required, but tails are not. Neither are spars!

**Q:** Okay, since there are many different kinds of kites do they fly differently?

**A:** Yes. Many kite styles fly much differently from each other.

**Examples:**

An Indian fighter kite is unstable while a delta kite is very stable. Different kites fly at different angles. Some pull less or more than average. Some kites are controlled by multiple lines while others flip over and over to fly.

**Experiment:** Try flying different kite styles. You might try these:

- single-line flat kite
- single-line bowed kite
- dual-line kite
- quad-line kite

After you fly them, write a summary to explain how they are different in flight.

A kite can be built from a wide variety of materials including paper, nylon, polyester, plastic, wood, carbon, and fiberglass. Some materials are common like plastic and paper. Others are less common.

Here is an overview of kite components:

**Eddy Kite**

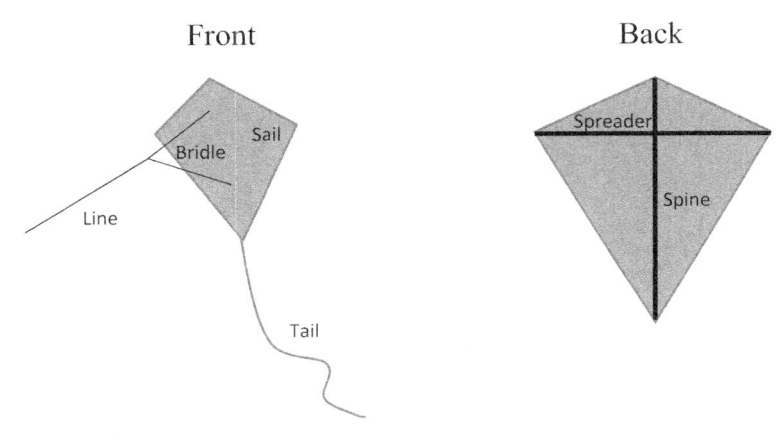

Front — Sail, Bridle, Line, Tail

Back — Spreader, Spine

**Delta Kite**

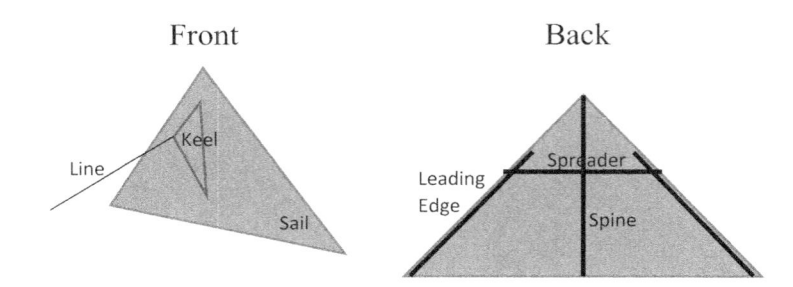

Front — Keel, Line, Sail

Back — Leading Edge, Spreader, Spine

Kite components include:
- Sail
- Frame and connectors
- Bridle
- Tails
- Flying line (the tether)

## Sail

The sail is surface on the kite that captures the wind. Kites may have a single surface such as bowed kites or multiple surfaces such as cellular kites. The kite surfaces should work together for best results.

### Deforming, flexing, or bagging

The kite will flex in the wind. What starts as a single surface on the ground may act as two different surfaces in flight.

### Sail size or "scale effect"

Kites scale well. The same kite style can be built in many different sizes. When the sail is larger, the kite will perform differently by capturing more wind and generating more lift with more drag. Therefore the larger version will pull more than a smaller version. As the frame gets larger the relative diameter of the frame may change the flexibility and stiffness of the kite.

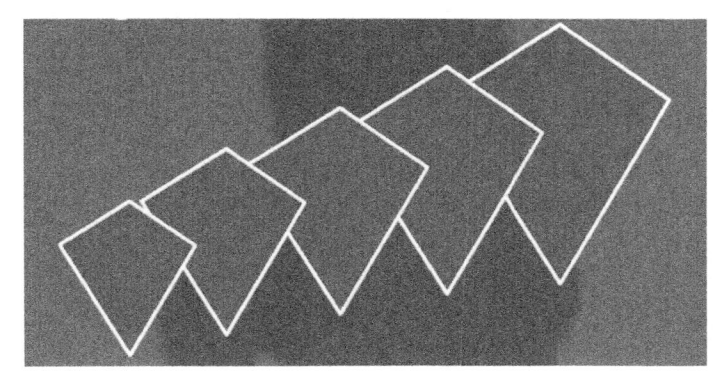

## Aspect Ratio

Some shapes fly better than others. The *aspect ratio* is a comparison of the height to the width.

**aspect_ratio = kite_width / kite_height**

How does it affect flight? In general, taller kites are better for high wind and wider kites are better for low wind.

**Experiment:** Measure and compare the height and width of a selection of kites to determine their aspect ratios.

## Sail loading

Sail loading is a comparison of the weight of the kite to the size of the kite sail. This comparison tells us when the kite is too heavy to fly and it also tells us how well a kite will fly in light wind.

Sail loading is the ratio of the weight to the size:

**sail_loading = kite_weight / kite_sail_area**

Because of its buoyancy, a hot air balloon is a great illustration of sail loading. Heating the air inside the balloon causes it to rise. Adding heavy sand bags causes it to fall. With a kite the wind does the lifting and the weight of the kite is what opposes the lift. These two forces move the kite until a hovering balance is reached.

**Experiment:** Find a thin plastic produce bag from the supermarket. Any lightweight sheet of plastic will work. Open it as much as possible then drop it and observe how slowly it drifts to the ground. Next, compress the bag into a small ball and drop it again. The difference is dramatic. The ball falls much faster because the weight has remained the same but the sail area is much less than before.

**Example:** Another way to look at sail loading is to think about a kite that flies well. When you attach more weight it will become too heavy to fly. It's the same kite but the sail loading has changed.

Lightweight and large kites fly better than heavy and small kites. If you have two similar kites that are:

- <u>Equal in weight</u> – the larger kite will fly better
- <u>Equal in size</u> – the lighter kite will fly better

**Experiment:** Calculate the sail loading for a few different styles of kites. Create a bar graph organized by weight.

## Altitude

**Q:** How high am I allowed to fly my kite?

**A:** Most kites should be flown below 500 feet in the USA. If your kite is under 5 pounds you can fly it above 500 feet. For safety, you must not fly kites within 5 miles of an airport. Verify these and check other safety considerations using the code of federal regulations 14CFR Part 101.

## Framework (with spars)

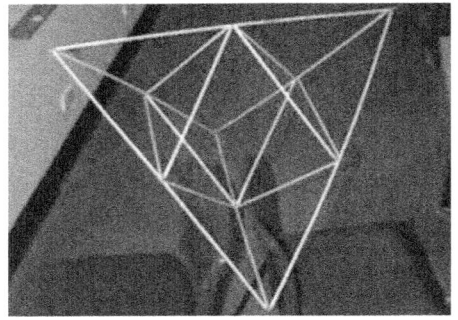

Modern kites frequently have a framework composed of rigid spars of carbon or fiberglass. To reduce the weight spars can be hollow tubes. A contributor to the strength of a tube is the thickness of the wall.

Many traditional kites are made with a bamboo frame.

In all cases the thickness of a spar is a large contributor to the stiffness of the spar but it also increases the weight. Shortening the spar will also make it stiffer.

**Dihedral and bow**

Some kites are flat, but many kites include a *dihedral angle*.

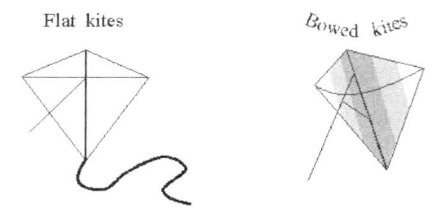

Flat kites       Bowed kites

A dihedral is a "v" shape that helps the kite to fly by diverting the air around the kite in the same way that a ship parts the water as it moves forward. Many kites use a 30 degree angle.

$30^0$

14

A *bowed* kite uses a curve instead of an angle to achieve the same results. Both of these kites are bowed using a curved horizontal spreader:

Spars can be used to:

1. Create a straight spine
2. Create a bow or dihedral
3. Lay flat against the kite to act as a *batten*. These are not under tension. They are added to keep a portion of the sail flat like the feathers on the wing of a bird kite.
4. Flex with the wind. In flight, these spars are thin enough to flex due to the force of the wind. This method is commonly used by Indian fighter kites.

15

There are soft kites such as sled kites and airfoil kites that are inflated by the wind during flight so they do not need spars. Instead they have additional bridle lines to distribute the stress across the sail and allow the kite maintain its shape.

This was the world's largest kite at 11,000 square feet but no frame.

## Bridles

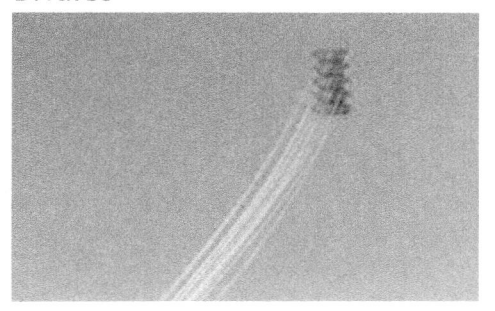

A kite bridle can range from a single line to more than 50 lines.

The bridle sets the angle of flight and supports the kite. Let's look at each of these functions of the bridle:

**Support the sail** – more bridle lines allow thinner bridle lines and fewer, thinner spars. Here is an example of bridle locations for a few different kites:

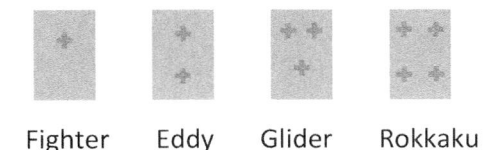

Fighter     Eddy     Glider     Rokkaku

**Set the angle of attack** – This is the angle that the kite is tilted into the wind. By making the top bridle lines shorter than the bottom bridle lines the kite will tilt forward in flight. The angle of flight is important because it places the kite in the best position for it to generate lift and stability. The proper angle depends entirely on the kite and the wind speed.

**Light wind angle of attack:**

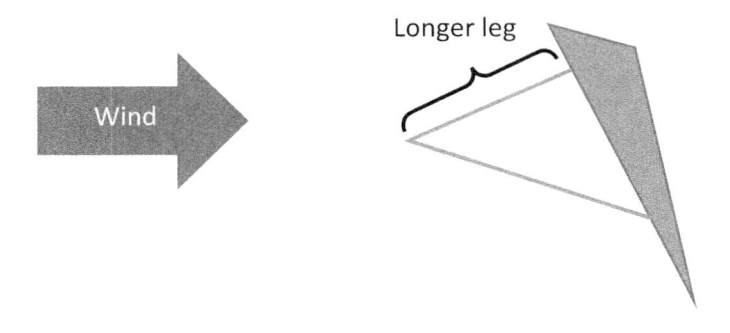

**Heavy wind angle of attack:**

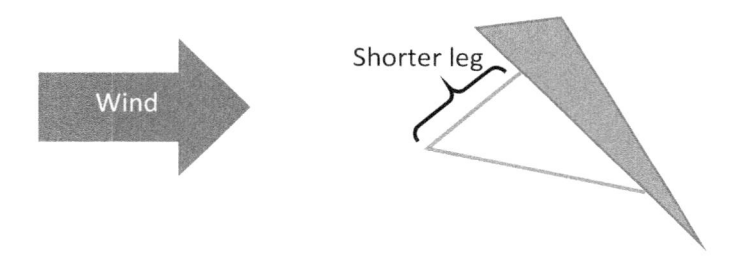

In heavy wind, the top leg of the bridle is shortened to change the angle of the kite into the wind. Tails are often added at the same time.

There is a point in front of the kite where the bridles come together this is called the *tow point.* The flying line connects the bridle point to the person flying the kite on the ground. That's called the *handle*.

To be considered as "flying," the tow point must be above the handle.

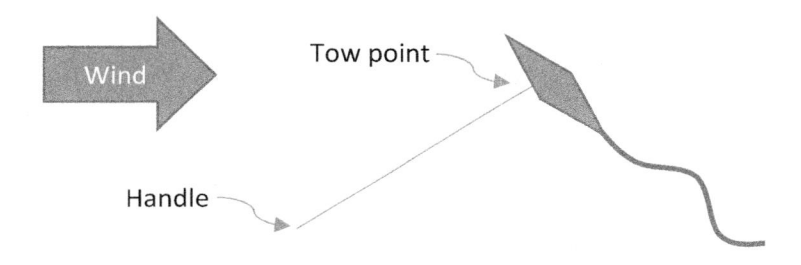

When the tow point is below the flying point it usually means that your kite is on the ground.

## Tails

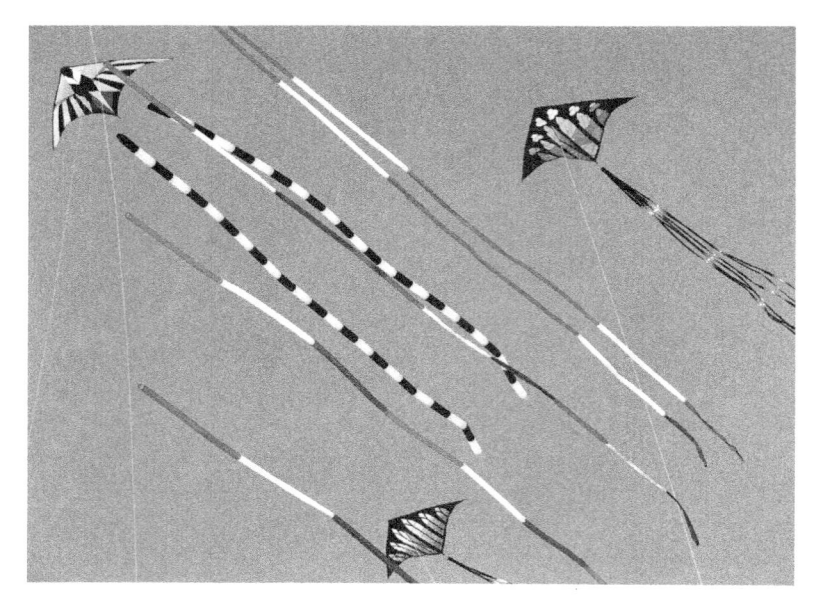

Kite tails can be beautiful, animated features. A long tail can provide additional stability to a kite that has no tail or a tail that is too short. Tails provide additional drag to:

1. Increase the stability of a kite in heavy wind conditions. Heavy wind can make a kite move left and right (yaw). Adding tails can reduce yaw during periods of heavy wind by forcing your kite to flex instead of turning left and right.

2. Point the kite into wind.

3. Add drag at specific locations.

4. Improve flight by keeping the kite at the same angle of attack without much added weight.

5. Make the kite look better.

## Tail Options

Kite tails can be wide or narrow, short or long, fuzzy, twisting, tube-shaped, numerous, U-shaped, Y-shaped, and many combinations.

**Examples:**

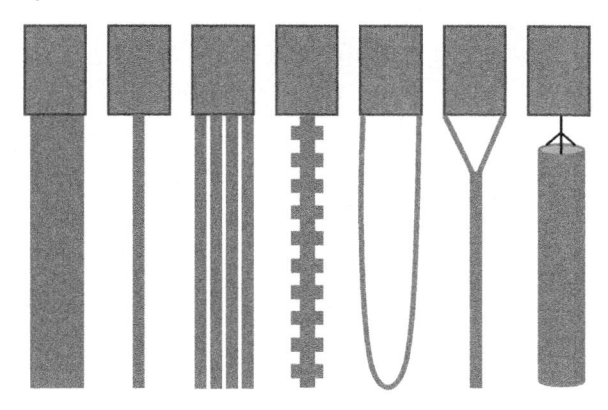

**Q:** How can we make a kite more stable?

**A:** Just add a tail. If your kite already has a tail, add more tails until the flight improves. A simple clip or a loop at the top of the tail makes it easy to add and remove tails as needed.

When you add tails to a kite, the general rule of thumb is to start with a streamer tail eight (8) times the height of your kite.

It's best to add tails at the bottom center of a kite in a symmetric manner. This will provide a greater stabilizing force because they act like a lever to keep the kite steady.

Fuzzy tails have cuts along their length and this makes them popular because they add more drag than a streamer of the same length and width.

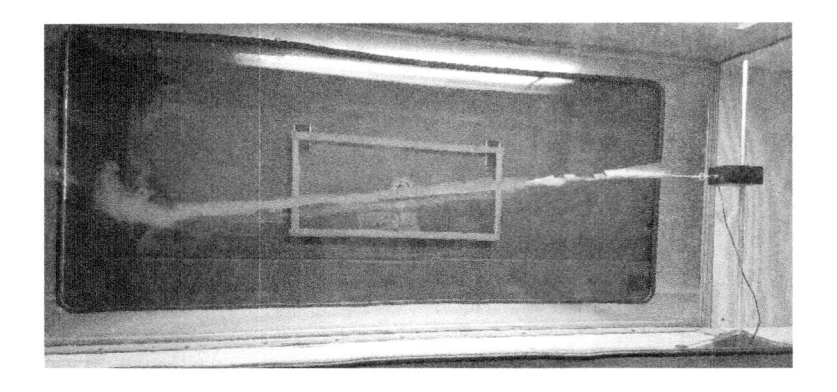

This is a fuzzy tail in a wind tunnel with a strain gauge to the right. Notice the short flapping section and the long straight section.

**Experiment:** Fly a kite with streamer tails. Double the number of tails to see the effect.

**Experiment:** Double the width of the tails.

**Experiment:** Double the length of the tails.

**Experiment:** Remove all tails.

**Q:** Which is best, a long tail, multiple tails, or a wide tail?

**A:** There's a widespread tendency to add a short ineffective tail as a decoration. Instead, longer tails are much more effective; perhaps eight (8) times the length of the kite. The length of the tails depend on the area of the kite, the wind speed, and the drag provided by the tail being added. Some kites do not need tails to fly well.

**Experiment:** Find out what happens when tails are too long.

**Experiment:** Find out what happens when there are too many tails.

> Why do kites have tails?
>
> So you can tell when they're happy!

**Windsocks, spinners, and drogues**

There are many ways to add drag to a kite. Most people think of adding tails but there are many other choices including windsocks, spinners, and drogues.

Let's take a look at each one.

## Windsocks

Many windsocks are decorative. Windsocks take many 3D forms and are often confused with kites because they can be attached to the flying line. They're not kites unless they generate lift and can fly on their own.

## Spinners

Spinners are similar to a propeller. They spin continuously in the wind. A fishing swivel will allow them to spin freely but what causes the spinning?

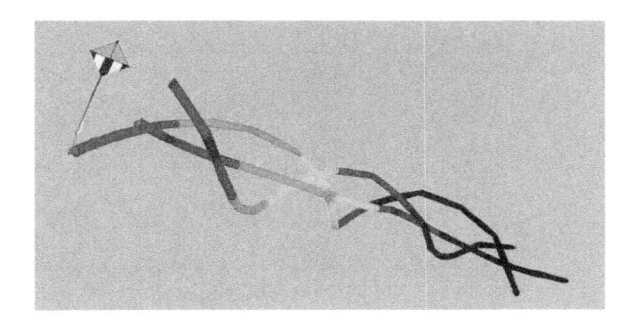

There are a few ways to enable spinners spin. You can have pockets tilted sideways, fins that are tilted like a pinwheel, or they might have an off-center design such as a helix or corkscrew. There are many different kinds of spinners that make dramatic kite tails because they are constantly in motion.

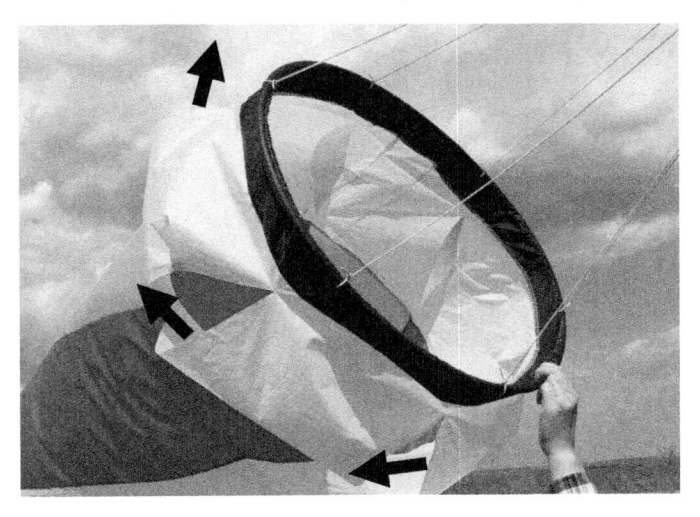

Air escapes to the side so this spins counter-clockwise.

## Drogues

Drogues are a great way to add drag behind a kite without long tails that can tangle with kites, lines, and trees. At the end of the day, it's easier to store your kite inside a drogue instead of collecting long tails. Drogues are like a bucket that trails behind a kite to catch the wind. They're interesting because they offer very little resistance at low wind speeds and much more resistance at high wind speeds. That's just what a kite needs.

Notice the taper from the mouth (on the left) to the back (on the right). More taper is necessary for larger kites that need the extra drag to make them stable.

**Experiment:** Replace a tail with a large or a small drogue. Which one flies best?

# Line (or string)

Kites couldn't fly without the line that keeps your kite from blowing away. The line has to be matched to the size of the kite and strong enough to handle gusts of wind. Kite line, string, and rope are available in a wide variety of strengths, diameters, lengths, and materials.

## **Line strength**

The strength of the line is rated by the breaking strength in pounds such as 10, 20, 50, 80, 100, 150, 200, 250, 500 pounds, and much more.

It is very important to match the strength of the kite line to the size of the kite and the speed of the wind. For most small kites 50# line is fine but it is possible to break 300# line using a large kite in heavy wind.

## **Line cutting**

Kite lines can be cut by the friction caused by one line rubbing against another.

When two kite lines cross during flight there's a good chance that one line will get cut. If neither line is moving you're likely to be safe. As soon as one line moves then the stationary line will be cut. The cut will happen faster if the two lines are under tension.

**Experiment:** Have two people hold two short pieces of line. Hold one line stationary. Move another line sharply against the other like a knife. Test various diameters and materials to see which one lasts the longest.

## Diameter

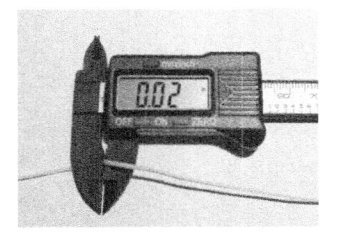

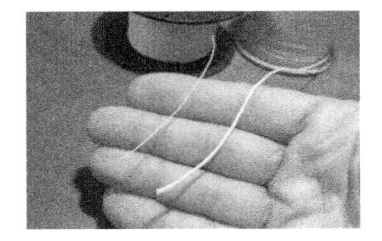

Digital Caliper                Thickness

Thicker line of the same material is stronger but it is also heavier. The weight and drag on the thicker line may cause the line to droop. This also happens when the wind speed decreases.

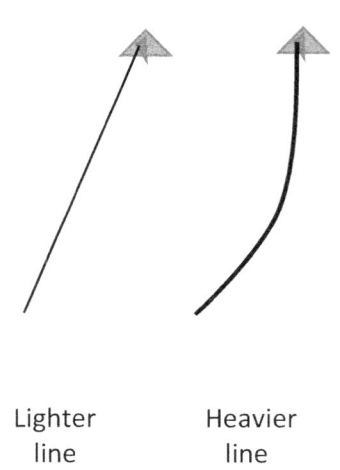

Lighter    Heavier
line       line

**Experiment:** Find multiple lines made from the same material, such as polyester. Compare the thickness of each line to the strength of the line.

**Experiment:** Find the breaking strength of the line by adding sand to an attached bucket. When the line snaps, weigh the bucket.

## Materials

Kite line is made from many materials including nylon, cotton, polyester, Spectra®, and various blends. Braided line is preferred over twisted line because it is less prone to knotting.

## Stretch

All kite lines stretch in an elastic manner. Like a rubber band, they stretch and return to their original length. This is different from stretching a plastic bag that remains stretched out.

Some lines stretch more than others and, as an example, nylon stretches more than polyester.

Stretching a kite line acts like a shock absorber for single line kites that are flying in gusty wind. Those who fly dual-line and quad-line kites prefer flying with kite lines made from Spectra® fiber because it has the least stretch and therefore it is more precise when using multiple lines to control the movements of a kite.

**Experiment:** Fly the same dual-line kite with identical length pairs of Spectra® line and nylon line. Do you notice any difference in the stretch?

## Length

Some people like to fly their kites into the clouds. A kite that is further away will look smaller. Remember that the higher you fly your kite the more line you will need to retrieve at the end of the day!

**Q:** Is the length of the flying line important?

**A:** Yes. Flying on a shorter line will keep the kite closer to you and:

- Allow others to see the size and design on the kite.
- Take less time to retrieve the kite.

Flying on a longer line will:

- Allow you to fly your kite in the steady air above trees and buildings.
- Give you more time to react when the wind drops and the kite starts falling. As the kite falls, you should react by pulling in line or moving upwind.

## Line vibration

When you're flying a kite and the wind is strong you may notice that the line begins to vibrate. This happens when the line is taut because the line acts like a very long guitar string being plucked by strong wind.

On the negative side, the line can pass vibration and swinging motion to their payloads such as a camera. This causes blurry photos.

**Experiment:** While the line is vibrating hold a card or a paper cup gently against the line to amplify the sound. Consider recording the sound at different wind speeds or after changing to a line with a different diameter. Does the sound change?

**Experiment:** With your kite high in the air, aggressively wave the spool of string up and down many feet at a time. Stop, then aggressively wave the spool of string left and right. Notice the movement of the wave as it travels up to the kite. This is one of the objectives in 8$^{th}$ Grade science to *"Describe the spread of energy away from an energy-producing source."* (8th Grade Core Curriculum)

The wave in the kite line is called a diminishing sine wave:

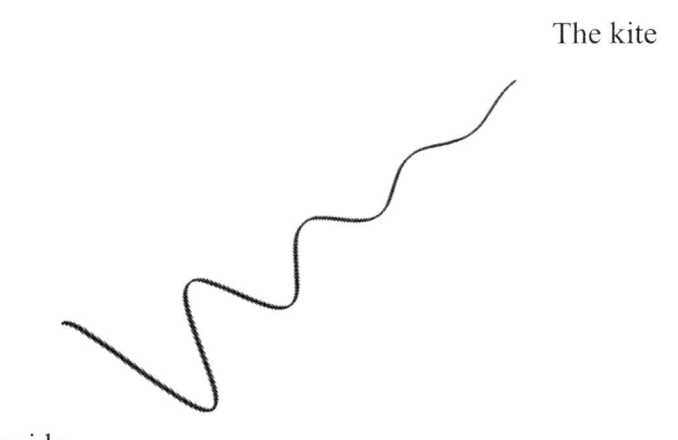

The kite

Your side

**Experiment:** Attach a ping pong ball to the line to allow you to clearly see the oscillations. Use a razor knife to cut half way through the ball. Slide the flying line through the cut. That should hold the ball in place without tape.

# Build a simple kite model

| Difficulty | Easy |
|---|---|
| **Time to build:** | 10 minutes |
| **Size:** | 4" (10 cm) tall x 3" (7 cm) wide |

This folded paper kite is easy, inexpensive, and the materials are easy to find. This design is perfect for classroom experiments.

This is one of the objectives in 8th Grade science to *"Design and build structures to support a load."*

## Materials:

- SAIL: Paper 3" x 4" (7 cm x 10 cm)
- SPARS: None
- TAIL: 3' (1 m) x 0.5" (1.5 cm) strips cut from a plastic bag
- LINE: Any sewing thread

## Tools:

- Pen
- Ruler
- Cellophane tape

## Method:

1. Fold paper in half

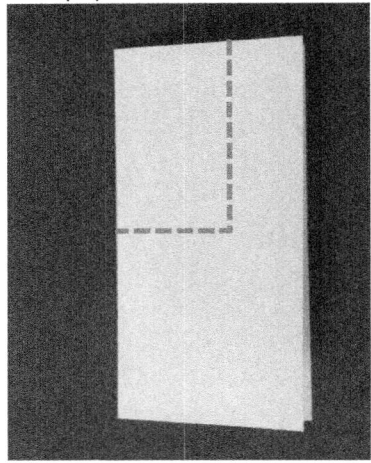

2. Cut to size 4" (10 cm) tall x 1.5" (4 cm) from the fold.

3. Unfold.

4. Tape the thread 1" (2.5 cm) from the top.

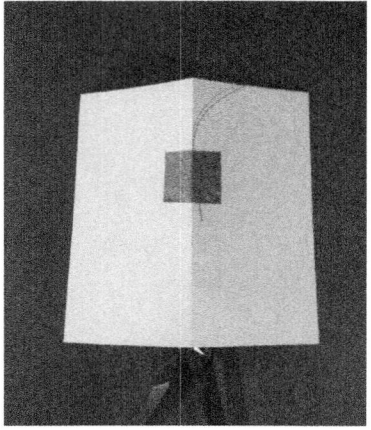

5. Tape the tail in the center at the bottom.

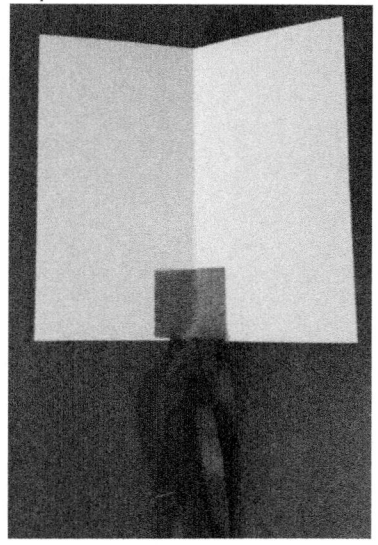

6. Test the kite by walking down a hallway. Test again by flying it too slow, just right, then too fast. This kite can be used to do a variety of experiments and exercises in this book.

**Experiment:** Design and build a kite. This is one of the objectives in 8th Grade science to *"Design and build a machine that uses gravity to accomplish a task."* (8th Grade Core Curriculum)

35

## Kite flying machine

A simple machine can be built to fly a small kite and do simple experiments. Let's call it a "kite flying machine."

**Q:** What is the difference between a kite flying machine and a wind tunnel?

**A:** The flying machine moves the kite through still air while a wind tunnel moves the air over a kite that stays in one place.

# Build a kite flying machine

**Materials:**

To build a machine to fly a small kite you will need a:

- Motor that turns at 16 rpm. These are available from www.scientificsonline.com
- Carbon rod with a diameter of 1/8" (3mm) and a length of 30" (75cm).
- Counterweight.
- Variable power supply for 3/6/9/12 volts.

**Method:**

1. Attach wires from the motor to the power supply.
2. Tape or use rubber tubing to attach the rod onto the shaft of the motor.
3. Attach a 5" (12cm) thread to a tiny, lightweight kite.
4. Attach the opposite end of the thread to the end of the rod so the kite dangles freely.
5. Attach a counterweight such as a pill bottle that allows you to add and remove weight.

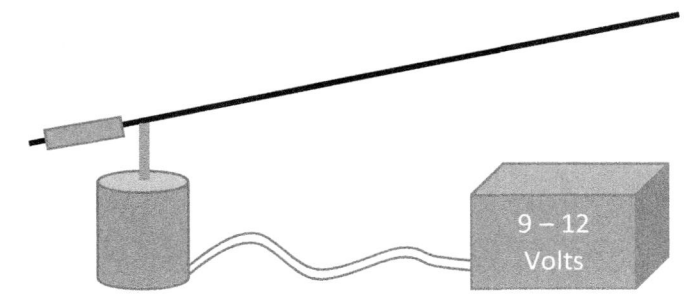

6. Slide the counterweight along the rod until the rod has the angle shown. You may need to add or remove weight.
7. Plug in the power supply.
8. Adjust the speed of the motor using the variable power supply:
   - Lower voltages, such as 3 volts, will show the reaction of the kite with too little wind. At that speed the kite will drag on the ground.
   - The kite should fly steadily at 9 volts. Make adjustments until it does.
   - Higher voltages, such as 12 volts, will show the reaction of the kite to too much wind. At that speed the kite will spin and crash.

**Experiment:** Build a kite flying machine using the instructions above. Build a simple kite model too. Attach it to the end of the rod using tape. Adjust the system to achieve steady flight. You may need to:

- shorten the thread
- add weight to the counterbalance
- change the speed of rotation by changing the voltage
- change the length of the rod (longer causes faster flight)

## The force of wind

From a gentle breeze to a blistery gale, wind is moving air. It's invisible but we know it's there because of the effect it has on the world around us. It moves sand, blows flags, pushes clouds, and causes the leaves on trees to dance.

Wind has direction, speed, and causes pressure against the things it hits.

The wind conditions encountered in the field include:

- Changes in wind speed (Very frequent)
- Changes in wind direction (Frequent)
- Thermals and updrafts (Rare)
- Ground turbulence that is caused by obstructions
- Surface wind that is different from the prevailing winds at higher altitudes

## What causes wind?

There are multiple sources of energy in the atmosphere and they can follow multiple paths that result in wind.

As the sun shines on the surfaces of the earth it heats some areas more than others. Warm areas cause the surrounding air to be less dense while cooler areas cause the air to be denser. The warm air rises so the heavier, cooler air moves horizontally to fill the void. This motion of air is called wind.

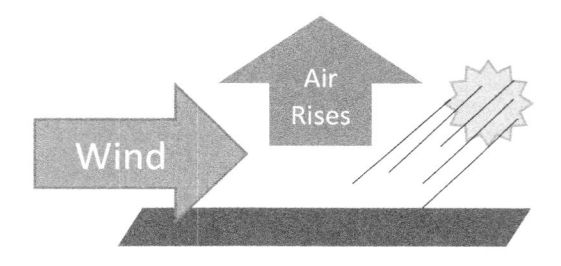

**Q:** Wind is moving air and that movement requires energy. Where does that energy come from?

**A:** Pressure variations, condensation of water vapor, and radiation from the sun. The sun provides a vast amount of solar energy that heats the air by first heating the ground.

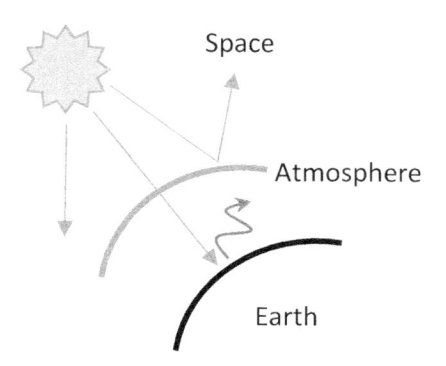

## Wind speed

In order to fly, kites need air moving over their surfaces. Some kites will fly in a wide range of wind speeds while other kites require lighter or heavier winds to fly smoothly.

Wind speeds generally change throughout the day. For that reason, many kite fliers carry multiple kites. This allows kite fliers to switch kites when the wind changes speed.

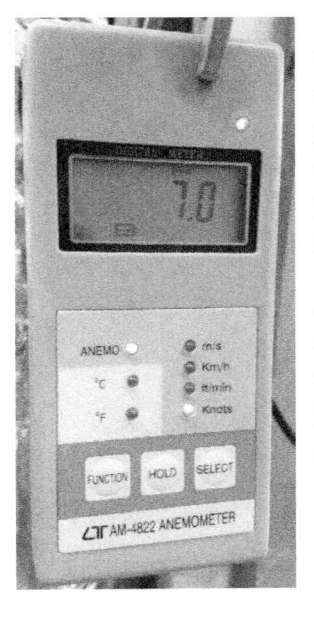

Winds also vary at different altitudes. According to Wikipedia, "Fluid flow is generally chaotic, and very small changes to shape and surface roughness can result in very different flows."

A digital *anemometer* can be used to measure wind speeds in a variety of units such as meters/second, kilometers/hour, feet/minute, knots, and m.p.h.

**Experiment:** Build an anemometer from paper cups, glue, and BBQ skewers. Use a thin nail at the center to allow it to spin freely. Mark one cup so it's easy to count the number of rotations done in 30 seconds. One student runs the stopwatch, the other counts the rotations.

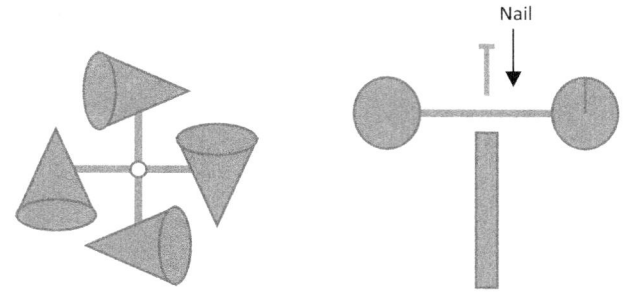

**Experiment:** Attach an anemometer to your kite line to measure the wind at various altitudes.

## Wind speed guide

| MPH | KPH | Observations | Example kites |
|---|---|---|---|
| 0-5 | 0-8 | At 0 mph smoke rises vertically. At 4mph wind is felt on your face. Leaves move. Flags begin moving. | Ultralight indoor kites |
| 5-10 | 8-16 | Branches move. Flags extend. | Most kites will fly |
| 10-15 | 16-24 | Flags flapping. | Most kites will fly with tails added |
| 15-20 | 24-32 | Flags flapping hard. | Box, sled, and sport kites |
| 20-25 | 32-40 | Large branches move. Flags flapping very hard. | Vented sport kites |

42

**Reacting to low wind**

When there's not enough wind your kite will fall to the ground. To counter this, walking upwind or pulling in some line increases the pressure of wind against the kite. This results in more pressure on the kite sail and a tighter flying line.

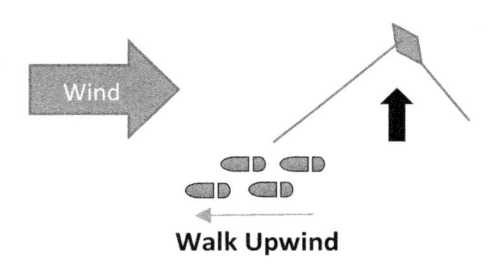

**Walk Upwind**

**Reacting to high wind**

Walking downwind results in less pressure on the kite sail and a looser line. Releasing some line has also decreases the pressure of the wind against the kite.

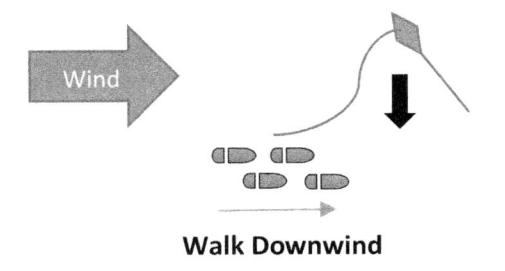

**Walk Downwind**

**Experiment:** While flying a kite, walk upwind and downwind. Try running upwind and downwind. What do you notice?

**Experiment:** Chart the wind speed over time using an anemometer. Each morning and afternoon measure the wind at the same location and at the same time of day.

**Q:** How does a kite react to the wind?

**A:** In light wind the kite may not fly. In moderate wind the kite may fly steadily. In heavy wind most kites become more active by swaying back and forth exhibiting a "yawing" motion. In addition, the kite spars may bend and the sail may change shape from flat to a curve allowing the kite to spill wind from the sail and remain stable. Kites that don't change shape may spin or crash and require longer tails to remain flying.

## Wind range

Most kites have a wind range from 5mph to 15mph (8kph to 24kph) and there are specialty kites that will fly below or above that range.

| 0 - 4 | 5 - 15 | 15 - 30 | + |
|---|---|---|---|

There are indoor kites that fly while you walk and can fly in a wind range from 0-4mph.

There are specialty kites that will fly up to 30mph (48kph) and very few kites are able to tolerate any winds that are above that speed. Higher speeds deform kites. This may force them into a shape that is no longer aerodynamic so they crash.

Gusting wind also creates problems because kites must be able to remain flying as the wind transitions from slower to faster to slower again.

In theory, larger kites will have a wider wind range but that's not always true because there are many factors involved including the material strength and the kite's angle of attack to the wind.

## Best winds

The best winds are steady winds with wind speeds that remain nearly the same. This allows you to select a kite, adjust it, then have it fly all day long.

## Wind direction

> "Kites rise highest against the wind, not with it."
>
> - Winston Churchill

Wind direction is determined by the source of the wind. If we say "It's a northerly wind" that means that we must face north for the wind to hit us.

Surface winds at ground level will follow complex patterns around objects such as buildings, hills, and trees. Inexperienced kite fliers expect the wind to be the same everywhere while experienced fliers know the importance of open access to wind before launching their kite. This requires that you take your kite to a better location in order to gain clear access to the wind.

Geography is also important for wind direction. The mountains, hills, and valleys will interact with wind at many flight sites. This is important for getting off the ground, especially in an urban river valley such as Portland, Oregon.

Those who fly in hot air balloons know that the wind pushes their balloons. They also know that as they go higher the direction of the wind may vary by a few degrees.

**Experiment:** Build a weather vane to indicate the wind direction.

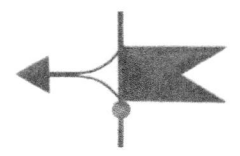

**Experiment:** Determine the wind direction using a flag, banner, streamer, or windsock. Use a compass to find north.

**Experiment:** Identify a few additional ways to detect the wind direction. There are many others.

**Q:** Can a kite be used to indicate the wind direction?

**A:** Yes. That's easy. The kite is a giant pointer in the sky telling you, "Hey! The wind is pointed this way!" Flying your kite higher allows you to gather data about the wind direction at different altitudes. You'll need plenty of string.

**Experiment:** Does a kite indicate the same direction at an altitude of 500 feet (150 meters) as the wind direction on the ground?

## Wind turbulence

Wind is generally chaotic. There is frequently turbulent air at the ground level that makes it more difficult to launch a kite. To understand this, think of the smooth motion of water at sea vs. tumbling water at the shore. Near the ground it gets messy.

Because of ground turbulence expert kite fliers use a "long launch" to quickly raise the kite above the trees and buildings into smoother, stronger wind.

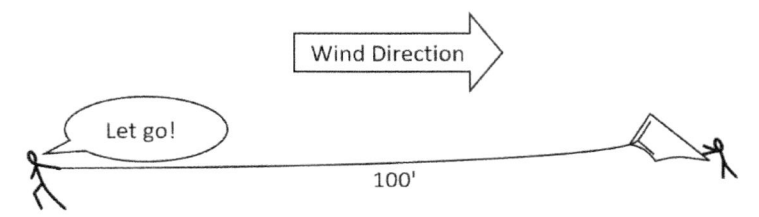

Throwing a kite is not a good way to launch a kite. It creates turbulence by having the kite move too fast and frequently in the wrong direction.

**Q:** What is the best wind for kites?

**A:** Most people prefer smooth, steady wind. This is called "laminar flow." Constant, smooth wind is usually preferred over any particular wind speed.

**Experiment:** Arrange two fans facing each other and put a streamer in the middle. Change the speed of the fans and the distance from the streamer while observing its behavior.

**Experiment:** Place many stakes, or poles in the ground using a grid pattern. The poles don't need to be very tall. Just attach a streamer to the end of each one. Record a video of the turbulence on the grid when the stakes are close together then again when they are far apart. Try moving the grid to interesting locations like near the corner of a building.

Photo taken in Cervia, Italy by Florian Janich of Germany.

## Wind shadow

Buildings and other structures cause turbulence. The turbulence behind a building creates a "shadow" that extends away from the building by approximately seven times the height of the building. In the shadow is the turbulent wind or dead air that can make kite flying difficult.

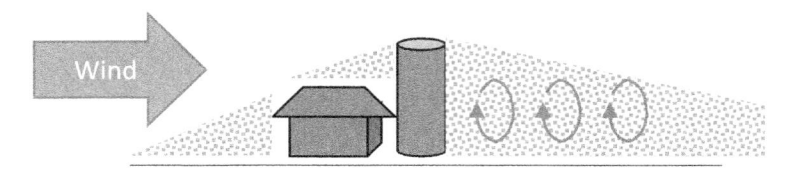

Sometimes the suitability of a certain location depends upon the wind direction.

### Trees

A baseball or soccer field might seem ideal until you notice that it's surrounded by trees.

To launch a kite, the wind direction will determine if we should stand to the left or right of this cluster of trees:

**Experiment:** Fans create turbulence so they're not a good way to fly a kite. If you have access to a wind tunnel you're in luck. Do some experiments with a kite and a tail. Predict what will happen when you change the size and shape of the kite. Take measurements then form conclusions.

## Forces on a kite

The forces on a kite can be measured as the average of its surface elements. What does that mean? It means that a kite can have many parts and you can measure them separately then put the information together to understand the big picture. The forces on a kite are similar to the forces on an airplane including: thrust, drag, lift, and gravity. Let's look at each concept separately.

## Concept: Thrust

Thrust is what pushes an object forward. On a rocket thrust is accomplished by pushing the rocket upward. On a kite, thrust is accomplished by using a line to pull the kite as the wind moves past it. This method is also used in water skiing, kite skiing, and kite surfing.

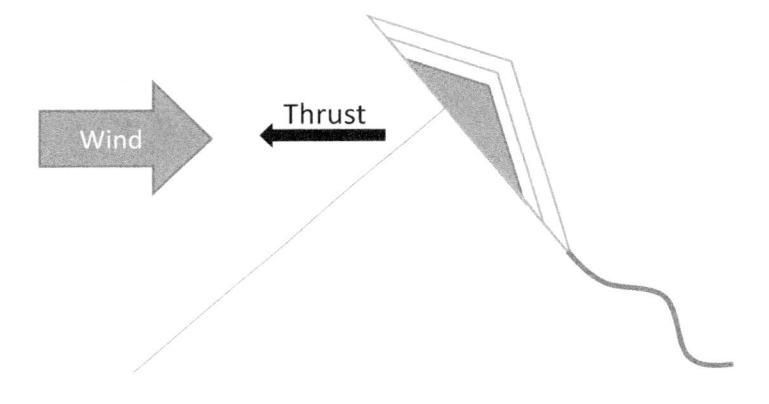

**Experiment:** Hold on to a rope and imagine yourself flying in the air as a kite. What happens when you turn your head, twist your shoulders, or move your arms and legs? Which direction would you move for each?

**Experiment:** Watch a video of a tethered sport such as water skiing, kite skiing, kite surfing, or ride a rope-tow up a ski mountain.

## Concept: Drag

Whenever wind hits an object there's drag involved.

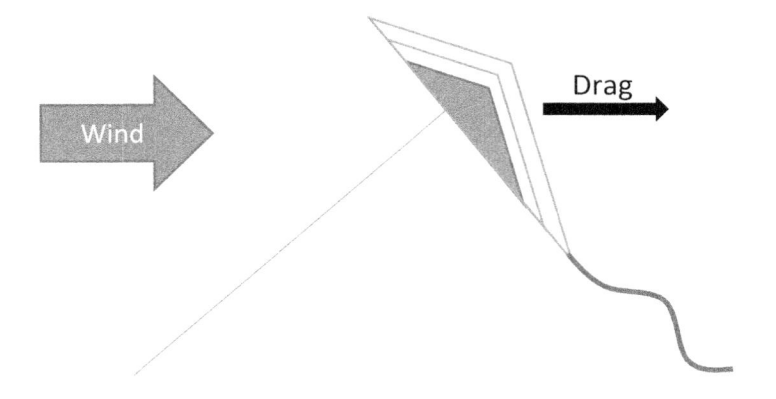

There are two types of drag that dominate kite flight. "Form drag" which is like hitting a wall and having to stop and walk around it while "Skin friction" is like pulling a blanket across sandpaper. Both slow you down.

| Form drag | Skin friction | Example |
|---|---|---|
| 0% | 100% | |
| ~10% | ~90% | |
| ~90% | ~10% | |
| 100% | 0% | |

**Experiment:** While riding as a passenger in a car, put your hand safely outside the window to feel the drag on your hand. Rotate your hand 0 degrees, 45 degrees, and 90 degrees to feel the difference in the drag. Did you also notice lift?

**Experiment:** Research the different types of drag.

Form drag is important for kites and is especially useful when we think about kite tails. Form drag has two components: steady and flapping.

**Examples:**

| | |
|---|---|
| **Drogues** | Provide form drag because they capture wind like a bucket |
| **Tube tails** | Provide skin friction |
| **Spin socks** | Provide form drag and skin friction |
| **Flapping tails** | Provide form drag at the end of the tail where they are flapping the fastest |

**Q:** Why do airplanes retract their wheels?

**A:** To save fuel by reducing drag for the entire flight.

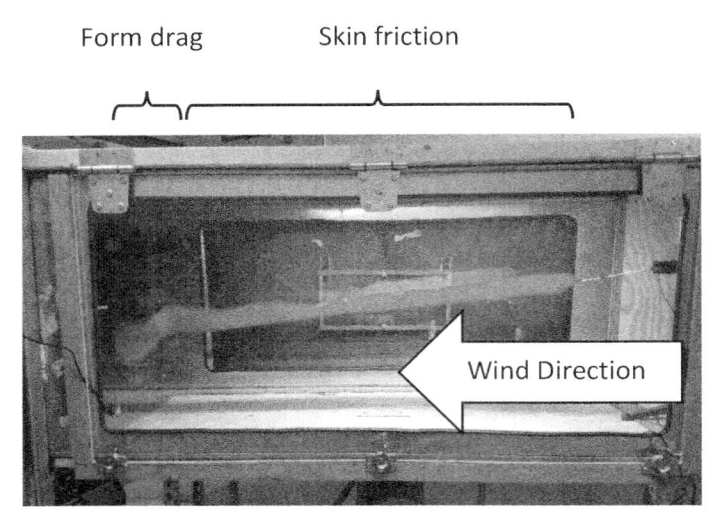

Here is a streamer tail in a wind tunnel.

Skin friction is a small component of the drag on a kite because only 1 to 5 percent of the drag of a tail is composed of skin friction. A long tail will have skin friction along the top of the tail with form drag where the tail is flapping at the bottom.

**Q:** What are a few of the items we attach to a kite to add drag?

A: Streamer tails, tube tails, fluffy tails, drogues, and flag/banner tails. Drogues increase their drag with increased wind.

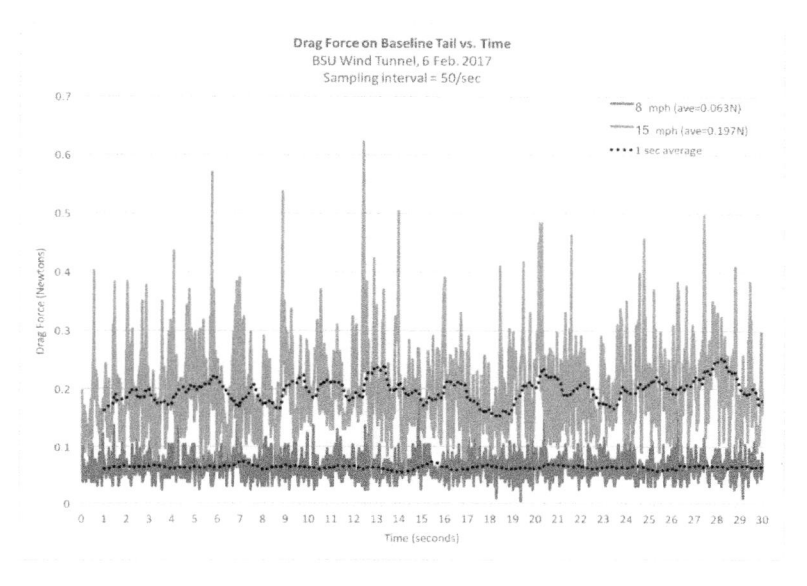

The drag force at 8 mph is less than drag force at 15 mph.

**Q:** Where do we add drag to a kite and why?

**A:** Usually we attach tails to the center of the bottom of a kite. Why? Because it's fun. Sometimes that's really the answer because some kites fly well without a tail so they don't need the additional drag. But attaching tails in-line with the spine is best because it keeps the kite straight and pointed into the wind. Matched tails can be attached in pairs on the left and right of the bottom center.

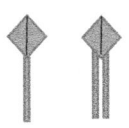

**Experiment:** Build a kite that uses drag to fly. This is one of the objectives in 8th Grade science to *"Engineer a device that uses friction to control the motion of an object."* (8th Grade Core Curriculum)

57

Here is an example of a kite that is flying well.

Adding a windsock to the line is fine but in this illustration we have added one that is too large:

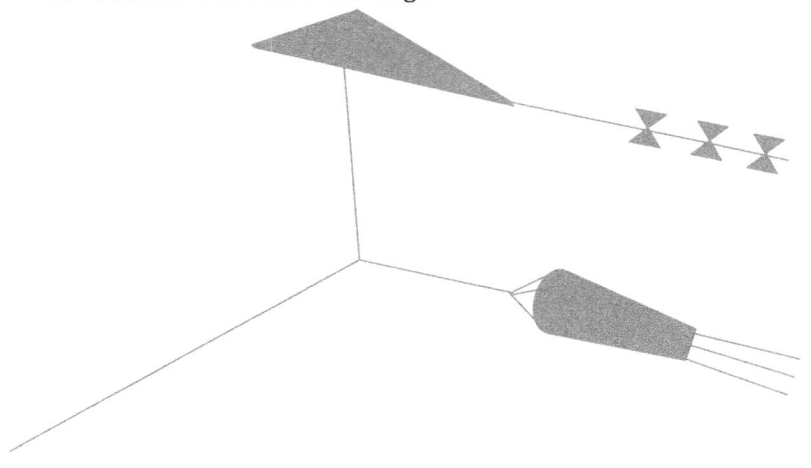

**Result:** The extra drag on the line has changed the angle of attack and may cause the kite to overfly or crash.

## Concept: Roll, pitch, and yaw

When you're flying a kite there are days when your kite may rise into the great blue sky then hang perfectly still. That's unusual. Usually there's some motion of the kite caused by the wind or by some property of the kite. In fact, kites fly using the same three-dimensional behavior as airplanes because they exhibit *roll, pitch,* and *yaw*. Here is a diagram of each one in order:

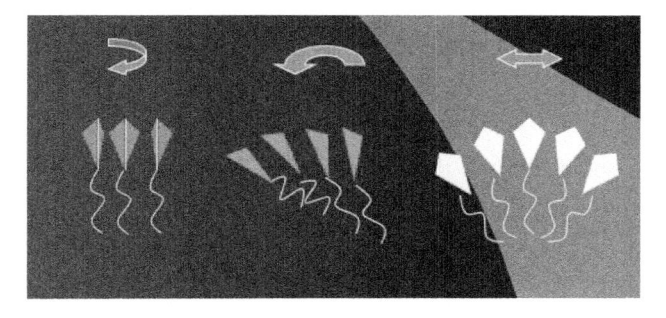

Here they are shown together on the same kite:

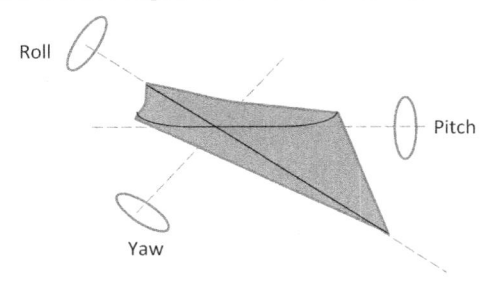

**Experiment:** Make a change to your kite and predict its effect on the flight of the kite. It is recommended that you make one change at a time. Changes may include:

- Making large holes on one side of the kite.
- Moving the tail.
- Replacing a spar with a thinner or thicker spar.
- Taping a quarter to one side of the kite.

## Concept: Lift

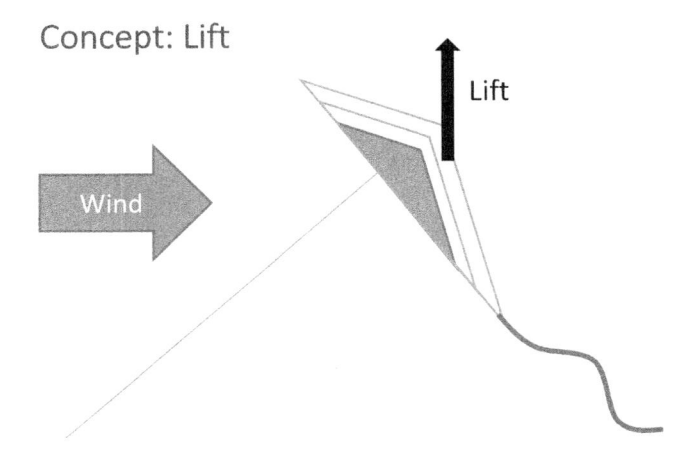

When people think of kites they often think about wind lifting their kites into the sky. Some people think lift is created by an updraft. Typically that is _not_ the case. Wind is usually horizontal to the surface of the earth and lift is the result of the kite's aerodynamic shape being tilted into the wind. The air molecules rush down along the kite acting as an invisible hand to lift the kite higher.

Let's use a Delta kite as an example:

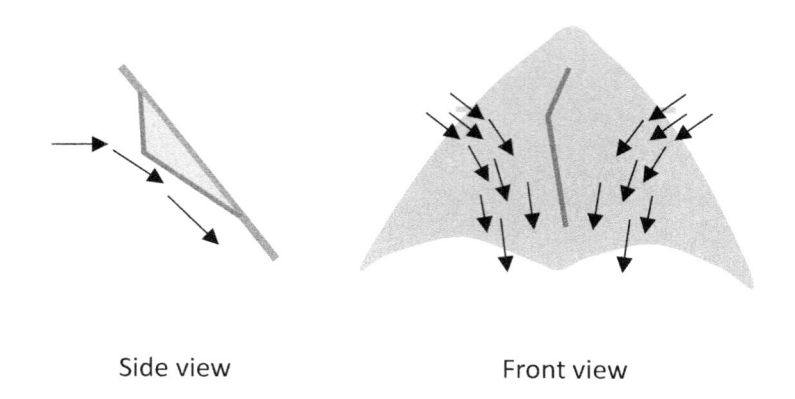

Side view                Front view

**Q:** What are some typical kite shapes?

**A:** There are many classic and traditional stable kite designs. Here are a few:

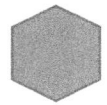

**Q:** Are there shapes that have poor aerodynamic qualities?

**A:** Yes, there are many of those too. They include: solid spheres, closed cylinders, solid pyramids, and many others.

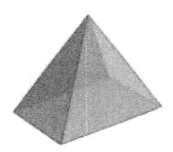

### Area

The lift is related to the kite's surface area. When flown at the same angle, larger kites catch more air and generate more lift.

**Q:** Are there multiple explanations for how kites create lift?

**A:** Yes, Newtonian lift and Bernoulli lift are two ways of explaining the lift created by kites.

## How Newton explains lift

As the air molecules strike the surface of the kite they are deflected downward. This action lifts the kite upward. This action lifts the kite because of what Newton called, "an equal & opposite reaction." This reaction is Newton's Third Law of Motion.

The wind is directed downward so the kite is pushed up.

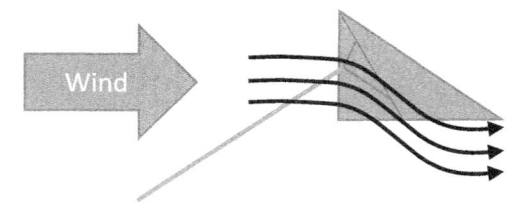

**Experiment:** Develop a model that demonstrates Newton's third law involving the motion of two colliding objects.

**Q:** How do box kites generate lift?

**A:** Box kites have multiple aligned surfaces that work together for the best results. If we treat a simple box kite as two separate kites we can imagine the wind lifting each:

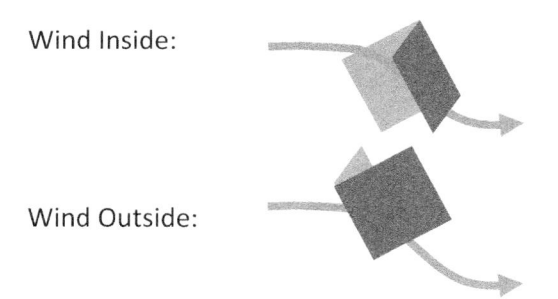

Wind Inside:

Wind Outside:

Lift is created by kite surfaces that work together. The surfaces could work against each other if they're not carefully aligned.

Some people have tried (perhaps accidentally) to build and fly a kite with misaligned surfaces. Since they are misaligned they fight each other to gain lift in different directions or block the airflow altogether. The result of those experiments keep the intended airship on the ground.

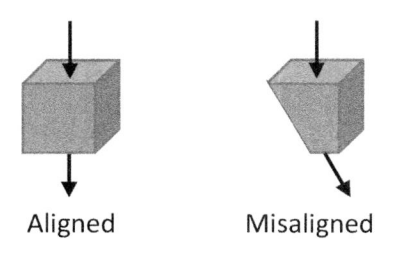

Aligned        Misaligned

**Example**: One day a little boy at Devereux Beach in Marblehead, Massachusetts built a typical Eddy kite in a workshop. He saw that it flew well, and proceeded to attach multiple Eddy kites into the shape of a ball. The new structure had surfaces that fought against each other so it did not fly.

## How the Bernoulli principle explains lift

While a kite is flying there is air moving past the kite. As the air hits the kite, some air moves above and some below the surface of the kite.

Because the pressure on the bottom is higher than the pressure on the top the result causes the kite to go up.

Here's a side view illustration of a kite with a flat surface that flies at a high angle:

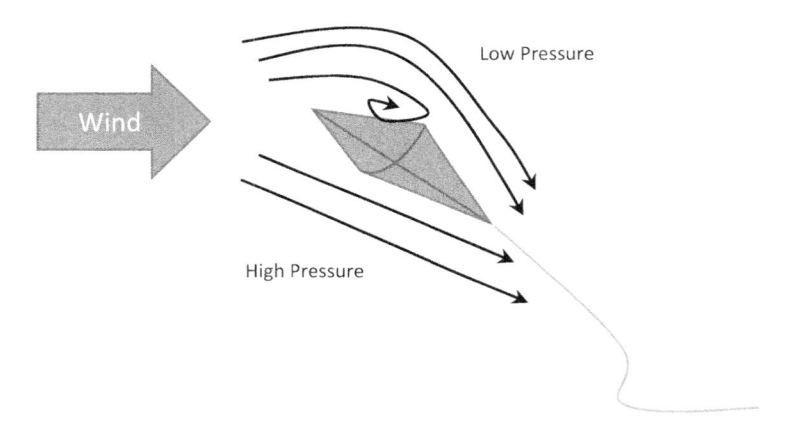

Notice that there is smooth airflow in front and behind the kite as well as turbulence behind the kite. The airflow is affected by the angle of flight.

**Example:**

Airfoil kites are great fliers. Notice the openings at the top of the guitar kite that allow air to fill the kite resulting in an airfoil that creates lift:

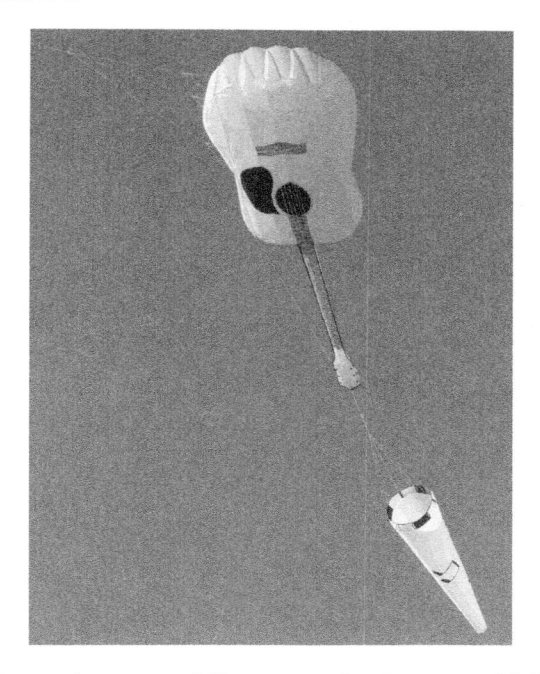

**Q:** Do different kites use different methods to gain lift?

**A:** Yes, there are multiple methods to gain lift. There are single surface kites such as dihedral or bowed kites. There are airfoil kites. There are multiple surface kites such as box kites, and there are multi-line kites.

**Experiment:** Hold a sheet of paper between your thumb and forefinger of each hand. Hold it about 4" (10 cm) down the 11" (27 cm) length. Notice how it hangs down. Blow over the top and see how the paper rises up toward the faster air you are creating.

## Adding lift by adding kites

Greater lift is accomplished by adding more sail area and adjusting the angle of attack.

One way to add more sail area is to add additional kites to your flying line. Make sure your flying line is strong enough to handle the additional tension.

*Kite arches* work by adding kites to the same line horizontally. Each kite that's added increases the amount of sail area and also adds lift.

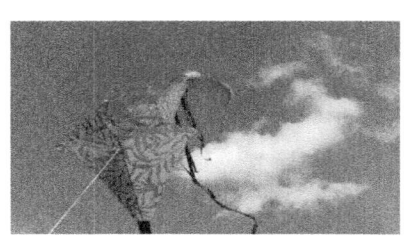

*Kite trains* work by attaching kites to the same line from front to back. This photo shows 50 kites in one train.

> **Did you know?**
>
> In 1895, Hargrave box kite trains were flown by the Blue Hill Meteorological Observatory in Milton, Massachusetts. They were used to lift instruments and do weather observations.

## A frequent misconception

Bouncers, bols, baskets, crowns, and parachutes are examples of devices that are called "ground bouncers" because the pressure of the wind causes them to stay open and bounce on the ground.

A pair of ground bouncers.

Some people expect them to fly but they are not kites. They generate some lift but they cannot sustain flight.

Similar to a parachute, a crown bounces on the ground.

## Lift factors

To summarize, lift is a combination of multiple factors working together:

- Area of the kite – larger area gives more lift
- Angle of attack – the proper angle increases lift
- Dihedral angle – a flatter kite gives more lift
- Wind speed – a higher wind speed generates more lift

## Concept: Gravity

Gravity is a law that cannot be broken. Everything that has mass will fall toward the nearest, heaviest object which is usually toward the center of the earth. Gravity is what brings a kite back to the ground.

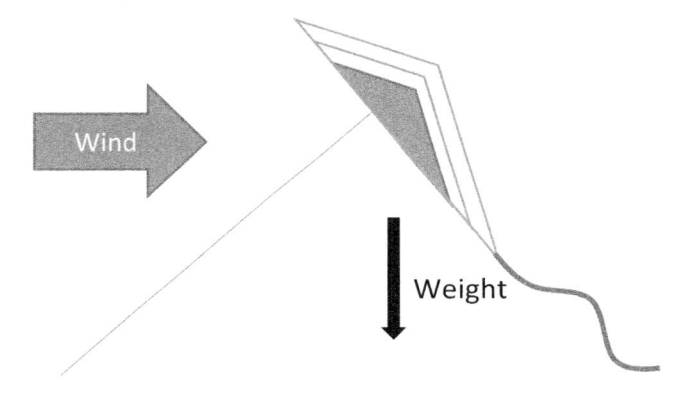

The weight of a kite will affect its flight because the wind creates lift but the weight of the kite is opposite that lift.

Some people think that a broken kite line would cause a kite to fly into outer space.

**Experiment:** This exercise requires two people. One person to fly a kite and cut the kite line while the kite is flying. Yes. Really. The other person should watch the kite from the side. Which direction does the kite fly? Up, out, or down? Why?

## Weight

When building a kite for the first time, some students give no thought to the weight of the kite. They believe that it needs extra strength to survive the wind and survive a crash but heavy kites or light winds will cause all kites to fall.

**Experiment:** Take a kite that you know flies well, then add weight to the kite or make an identical kite using heavier sail materials. Instead you could attach strips of masking tape, attach strips of duct tape, or attach coins to the original kite. Be sure the weight is added in a balanced manner at the center of gravity or around the bridle point. How does the flight performance change? This is one of the objectives in 8th Grade science to *"Cite examples of how Earth's gravitational force on an object depends upon the mass of the object." (8th Grade Core Curriculum)*

**Experiment:** Try to lift a payload. Launch your kite then 50' from your kite attach a payload to the line. If the kite can lift the payload add more weight. How much can your kite lift?

**Q:** Could we substitute weight in place of a tail? Would the weight be a good replacement to help a kite fly in heavy wind?

**A:** No, adding weight is not the same. Gravity is pulling on both the kite and the tail. An ideal tail should not add weight, it should only add drag. That way it's in line with the kite and does not change the *angle of attack* of the kite. The first illustration below shows the kite being pulled down and changing the angle of attack to the wind. Changing the angle of attack may cause the kite to crash. Having the tail *in line* with the kite is far better:

Poor:　　　　　　In line is better:

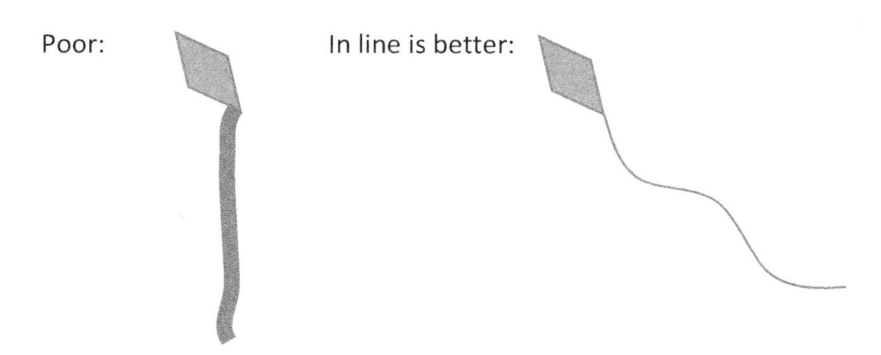

**Experiment:** Replace a kite's tail:

1. With a rope that weighs the same as the tail.
2. With a rope that's the same length as the original tail.

Does the rope fly just as well as the tail?

## Strength vs. weight

Kites are a *balance* of strength and weight. They must be strong enough to handle the force of the wind yet light enough to fly.

We avoid the heaviest materials (like metal) because they are so heavy. Maybe the strongest material isn't the best choice to form a balance.

Lighter kites fly better but lighter kites could break more easily. Again, balance is important.

## Materials

Two materials may look similar but one may be heavier than the other.

**Experiment:** Compare a variety of different materials. Cut 6" x 6" (15 cm x 15 cm) pieces of:

- plastic
- cloth
- Tyvek®
- copy paper
- tissue paper
- aluminum foil

Weigh each one. Tear them in half one at a time and add the results to the table. Is the lightest the easiest to tear? Is the heaviest the strongest? Form a table of results like this:

| Material | Weight | Easy to tear? |
|---|---|---|
| Plastic bag | | |
| Cloth | | |
| Tyvek® | | |
| Copy paper | | |
| Tissue paper | | |
| Aluminum foil | | |

## Concept: Stiffness vs. flexibility

Kites should be strong enough to survive the wind. Just as your body has a skeleton for support, the strength of a kite comes mostly from the kite frame.

Kite frames are usually built with spars that give the kite an aerodynamic shape that is designed to face the wind.

Spars can be made from a variety of materials. Some are thicker or thinner.

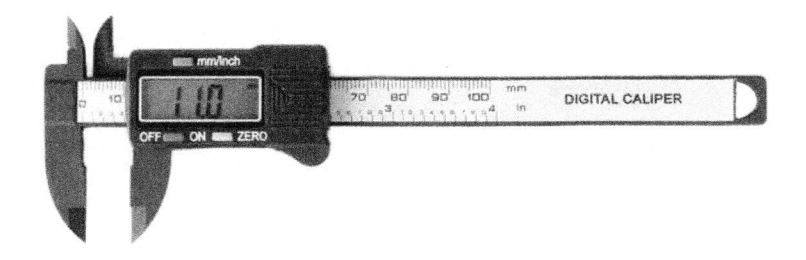

Calipers allow us to measure the thickness of spars.

A variety of diameters allows a kite frame to be stiff enough to maintain its shape or flexible enough to bend in the wind. Some of the rods below are hollow tubes to save weight while the thickness of the wall contributes to its flexibility.

**Experiment:** Hang a water bottle from the center of a spar and chart the deflection of different diameters of wood, fiberglass, and carbon fiber spars. This will show us how much it will bend in heavy wind. Notice that the length of the spar is important. Will longer spars bend more or less?

Q: Does the flexibility of a kite help it to fly?

A: Yes, The ability for a kite to flex is one method of stability. Flexing will allow the kite to adjust to the wind. More wind, more flex. As the kite flexes more it releases more pressure on the sail.

Here is a diamond kite in flight. Notice how narrow the bottom has become:

Here is a delta kite in flight. Notice how the sail billows in the breeze allowing air to flow on the left and right sides:

All kites flex but some kites use this property to their advantage. For example, the spreader on an Indian Fighter Kite is made to flex in flight. This makes the kite more stable than before. When you tug on the line the spreader flexes allowing the kite to zoom in the direction it's pointed.

75

Here is a flat fighter kite without wind:

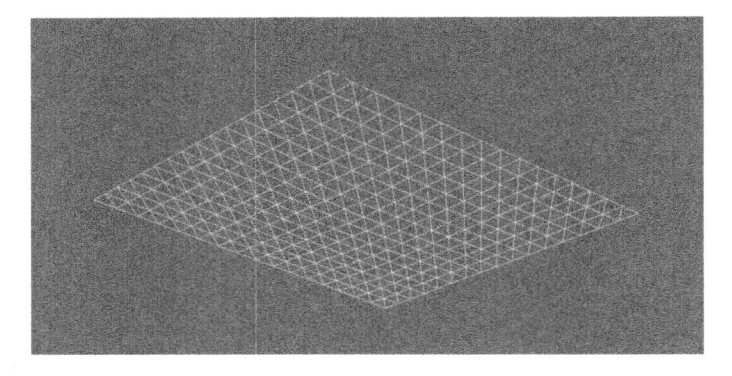

Here is the same kite when it is flexed by the wind:

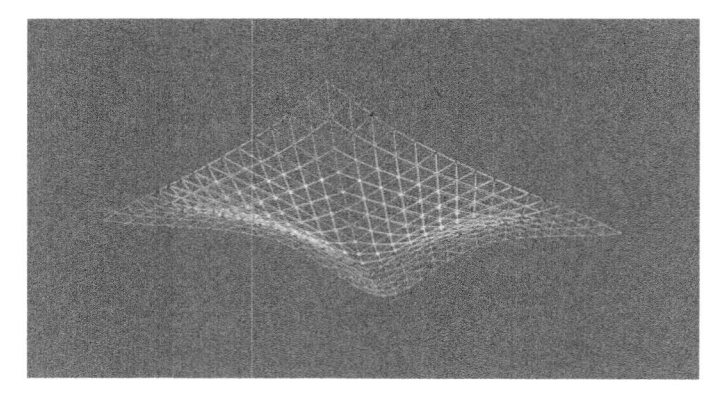

Notice how the center of the kite remains forward because of the spine in back, the bridle in front, and the pull of the line.

## Concept: Airflow over the kite

When a kite is symmetric, the left and right sides of the kite are identical. This provides balanced airflow over the surface of a kite.

As you are flying a kite, consider how the air moves across the surface of the kite. Here are some simplified examples but the air movement can be very complicated and it changes with the angle of attack.

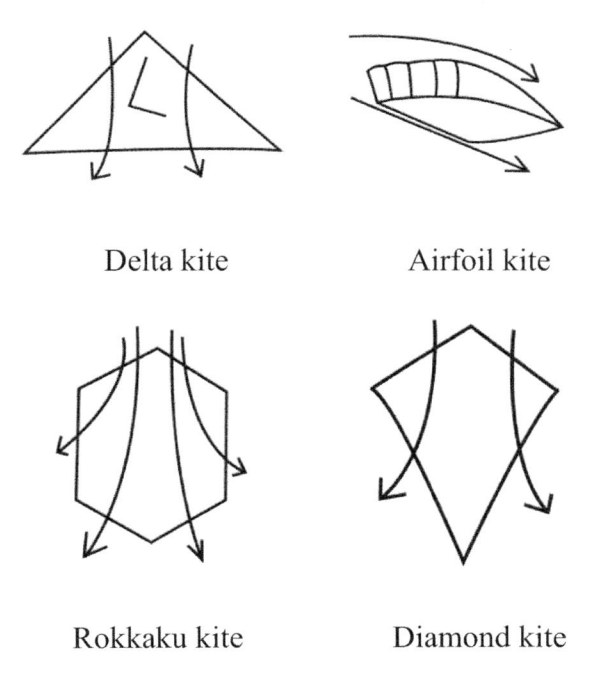

Delta kite

Airfoil kite

Rokkaku kite

Diamond kite

**Recommendation:** Don't throw the kite. The wind should flow smoothly over the kite to do the work of lifting the kite at the proper angle.

## Concept: Angle of attack

The angle of attack is the angle of the kite into the wind. It is measured against the horizon.

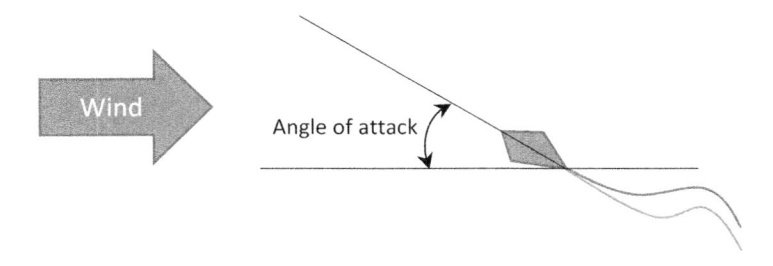

Each kite has an ideal angle that allows it to balance lift with stability. Tilting the kite more causes more wind to miss the kite. This is useful in heavy wind situations because it will reduce the pull on the line.

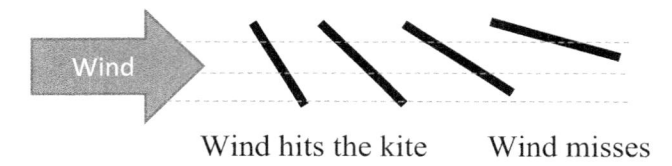

Wind hits the kite     Wind misses

**Experiment:** Use a protractor to measure the *angle of attack*. That's the tilt of the kite relative to the horizon. You may need to fly the kite close to the ground to take the measurement.

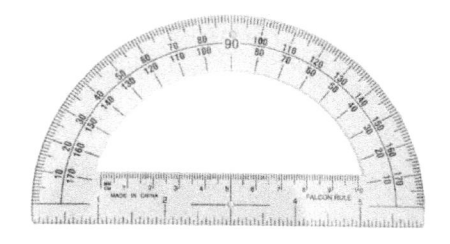

## Concept: Flying angle

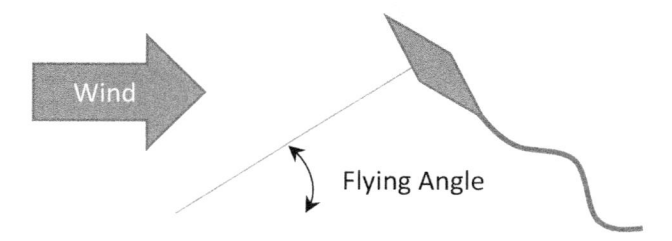

The *flying angle* of a kite is the angle of the line relative to the horizon. This can be different from the angle of attack.

**Experiment:** Use a protractor to measure the flying angle.

**Experiment:** Test fly a kite to make sure it flies well, then change the angle of attack by moving the bridle point <u>up</u> one half inch. Try flying the kite again. What happened to the flying angle?

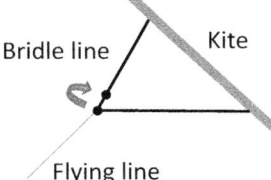

**Experiment:** Test fly a kite to make sure it flies well, then change the angle of attack by moving bridle point <u>down</u> one half inch. Try flying the kite again. What happened to the flying angle?

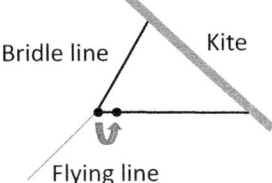

## Concept: Tension

Tension is the pull on the line. Tension is caused by the sum of the lift that pulls up and the drag that pulls down<u>wind</u> (not downward).

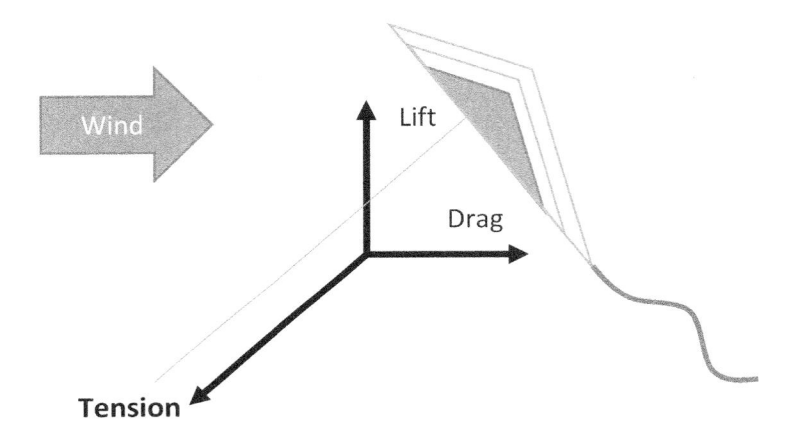

**Example:** Adding tails and decorations to the flying line does not add lift but it does add drag so the tension increases.

**Experiment:** Tie the end of three ropes together just as the arrows show in the diagram above. One person acts as "lift" and the other acts as "drag" by pulling at 90 degrees to each other. The third person will feel the *combined* tension in a third direction. Measure the angles to confirm the third direction.

**Experiment:** Measure the tension on the line of a kite using a luggage scale or a spring scale. You will feel the wind combined with the drag created by the kite, line, and tails. If you don't have a luggage scale, use a few heavy duty rubber bands and use a ruler to measure how much they stretch.

## Concept: Gliding vs. stalling

Gliding:

Gliding is when the aircraft is moving forward with its nose down and is losing altitude slowly. Airplanes, gliders, and many other types of aircraft do this.

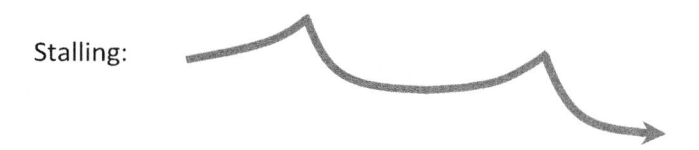

Stalling:

Stalling is a sudden reduction in lift when the nose of the aircraft rises causing the forward motion to slow or stop. This helps a kite to continue to fly when the wind slows or stops briefly. During a drop in the wind, the kite will lose altitude but the nose of the kite will rise again and again to try to catch the wind and recover.

The difference between gliding and stalling is partly based on the balance of the kite from front to back or "nose to tail."

**Q:** Do kites glide or stall?

**A:** Most kites will stall to bring the nose up again and again trying to recover flight when the wind drops. If the kite went into a glide it would not recover. It would glide back to the ground after the first drop in the wind.

**Experiment:** Build a glider from paper or balsa wood. Make changes to your construction so the glider can stall. Then make it glide again.

There is a kind of kite called a *glider kite* that will rise when you pull on the line. When the pulling stops the kite will begin to glide.

"Gliding kites have become a common indoor spectacle. With such a wide variety, each gliding kite has different flight characteristics, and some are so well behaved that anyone can fly them."  — John Barresi, *Kitelife.com*

## Concept: Inverse square law

**Q:** Why not fly your kite to the very end of the kite line?

**A:** Kites look tiny when they're a great distance away, but everything looks smaller at a distance, not just kites.

**Example:** "The sun is 1.4 million kilometers or 870,000 miles across. It's so big you could fit a million earths inside it. Yet you can block its light with your thumb."

There is an equation that relates the size of a kite that is nearby to one that is far away from our eye:

$$\frac{size\ of\ nearby\ kite}{size\ of\ kite\ in\ the\ sky} = \frac{distance\ from\ kite\ in\ the\ sky^2}{distance\ from\ nearby\ kite^2}$$

**Experiment:** A kite appears to be 1 meter tall when it is 2 meters away from our eye. When the kite is 100 meters away, how tall does it appear in the sky?

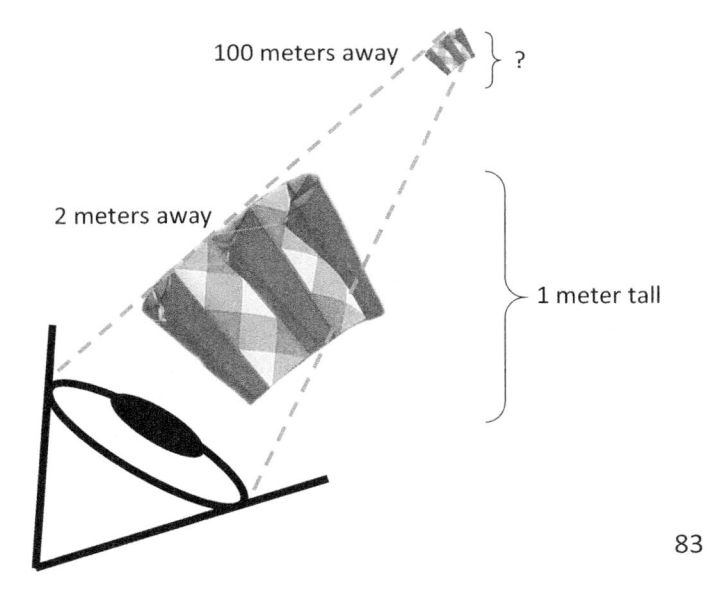

83

## How do I make my kite fly better?

Some kites fly better than others. That depends on many factors, not just the kite or the person flying the kite. The wind speed is critical. Successful kite fliers have multiple kites and choose one that is tuned to fly in the current wind conditions.

Once you've chosen a kite to fly, let us consider ways to make it fly better.

## Stability is important

A kite should be steady in flight. Many kites are so steady that they fly without interaction from the flier. The kite can be tied to a fence and no tugging is necessary! That's a stable kite.

> **Definition: Stability**
>
> The property of a body that causes it, when disturbed from a condition of equilibrium or steady motion, to develop forces or moments that restore the original condition.
>
> *—Miriam Webster Dictionary*

So the definition of stability tells us that a kite doesn't need to stay in one place to be stable. It can move around in the sky so long as it returns to equilibrium.

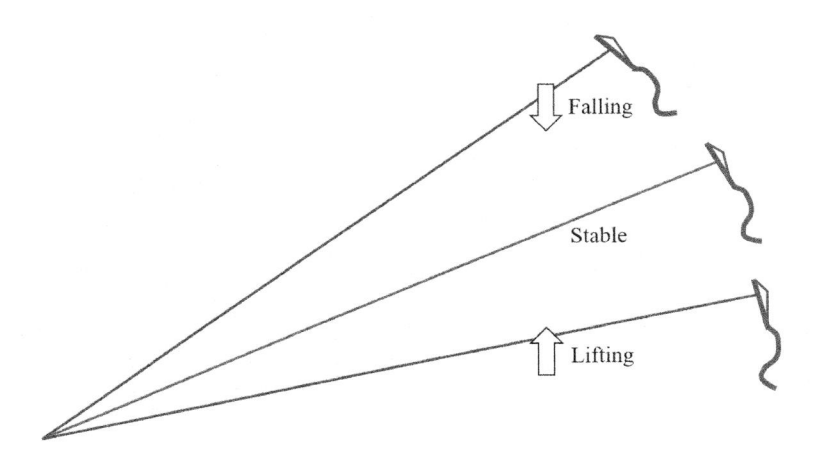

A kite is a self-equalizing system with the kite, line, and wind.

A kite must also be stable by returning to the center from the left and right sides.

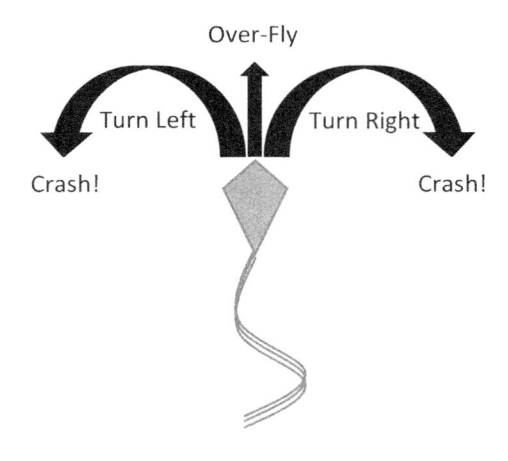

Over-Fly

Turn Left    Turn Right

Crash!                    Crash!

**Example:**

Let's consider what happens when a kite rolls around the longitudinal-axis. While in flight, that's like rolling around the center spine of the kite.

Let's look at this one step at a time.

First, the wind pushes on both wings the right wing lifts up and rolls the kite to the left. Once the right wing is up there is less pressure on the right wing than the left so the kite rolls back to equilibrium.

**Equilibrium:**

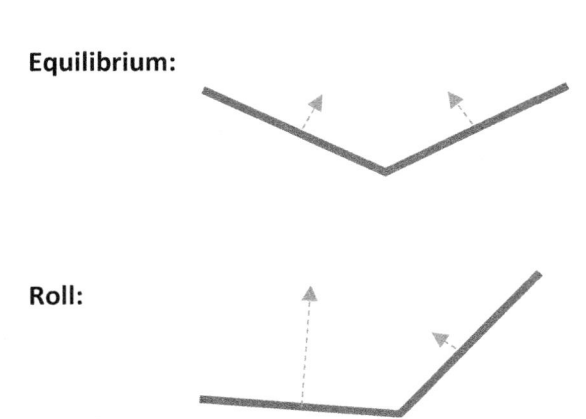

**Roll:**

**Back to Equilibrium:**

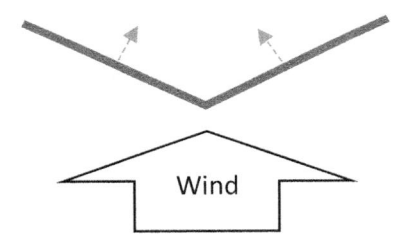

Similar actions take place for pitch and yaw.

**Q:** Why are box kites, Cody kites, and Rokkaku kites able to fly without tails?

**A:** Kites use many methods to remain stable in flight, not just tails.

## Methods to increase kite stability

These methods increase the stability of a kite:

1. Adjust the bridle to improve the angle of attack.

2. Add tails or a drogue to the bottom center.

3. Increase the dihedral or "bow" of the kite – notice the curve in this boy's sail:

4. Balance the weight of the kite left-right, forward-back.

5. Place balanced bridle attachment points. Consider adding additional bridle lines to support the sail and frame.

6. Use one of many aerodynamic shapes such as an Eddy, Delta, cellular kite, airfoil, or the Japanese Yakko sleeves.

7. Maintain perfect <u>symmetry</u>.

8. <u>Vent</u> the sail to allow air to pass through the sail thereby reducing pressure and giving less tension in heavy wind. This sail is vented and has spinners too:

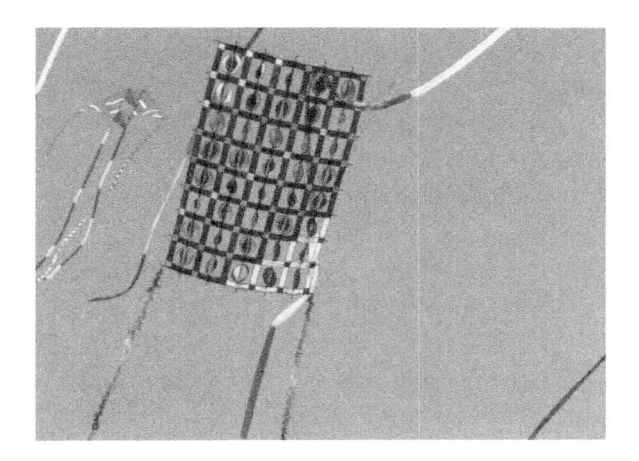

9. Add <u>keels</u> – the three keels on this kite are required for this great pair of legs to be able to fly:

10. Increase the <u>flexibility</u> by reducing the thickness of spreaders so the kite will flex in a wind gust instead of crashing. Fighter kites are well known for their precise movements while using flexibility to fly. Flexibility allows maneuvering and stability during wind gusts.

11. Use a low <u>aspect ratio</u> to increase stability.

12. <u>Stretch</u> is a part of flexibility that can increase stability. Bungee cords stretch to allow the flexing that prevents prevent breakage due to the shock of hitting the ground or the pressure of heavy, gusty wind.

13. Strengthen the <u>spine</u> to prevent it from being distorted by the wind.

14. <u>Reduce the weight</u> of the kite – this helps the kite fly in light wind. This can be done using hollow spars, lightweight and thin sail materials, removing metal connectors and swivels, and lightweight kite line.

These factors can be combined. For example, in heavy wind you might change the angle of attack *and* add tails. This combination is a common practice.

## Adjust the bow or dihedral angle

Once of the features of a kite that creates balance is the curve of the sail. This causes the wind to split onto both sides of the kite while the kite stays centered.

Without a curve or dihedral the kite may be unstable. Here is a top view of a kite with a flat spreader:

Flat: ———————

Kites are frequently bowed by placing a tension line across the wingtips or making the spreader thin enough so the wind causes it to bow during flight.

Bowed:

A dihedral is formed by using two spars that are joined in the center by a fitting that holds the spars and sets the dihedral angle.

Dihedral:

A dihedral can be reinforced with a third spar for added strength and stiffness but reduced flexibility.

Braced:

Here are a few examples of the fittings that form the dihedral angle. On the back of a kite the connector should be facing up. These are different colors because they accept different diameter rods.

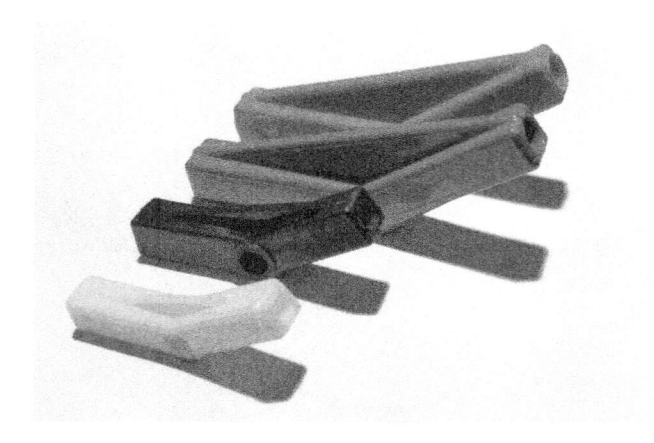

## Adjust for stability vs. lift

Think of stability and lift as two opposing ideas. Changes that add stability can reduce lift but changes that add lift can reduce stability. It seems like a conundrum doesn't it? Well, really it's not. We're simply looking for a happy middle based on the current wind speed. Some kites are made to be adjustable for exactly that reason.

One method is the increase the "bow" of the sail. We can create a kite with more bow by curving the spreader more:

**Q:** How does this change the flight?

**A:** It changes both the lift and the stability.

| Action | Lift | Stability | Result |
|---|---|---|---|
| <u>More</u> bow or dihedral angle | Less | More | The kite is steadier in flight |
| <u>Less</u> bow or dihedral angle | More | Less | The kite flies at a higher angle |

## Check the symmetry

Kites have a Zen nature to them. Kites need to have the balance that is formed when the left side is equal to the right side. That balance is formed by *symmetry* meaning "the same."

**Google definition:** "the quality of being made up of exactly similar parts facing each other or around an axis."

Here are some examples of popular shapes that are symmetric. These have been used many times to make kites:

Here is a shape that is not symmetric:

And spars that are not symmetric:

Right side is too long.

Left side is too bent.

**Experiment:** Some interesting kites have been created that are not symmetric but continue to fly steadily. Try to create one yourself.

For the best flight, kites need these properties to be symmetric:

| Size | The left side should be equal in *area* to the right side. |
|---|---|
| Shape | The shape of both sides should be a *mirror image* of each other. |
| Weight | The weight should be *balanced* on the left and right. |
| Strength | The two sides should *flex* the same way. That means that the spars should be identical in thickness and length. |
| Bridle | Some kite bridles have many lines. The length of the lines must be equal on the left and right. |
| Tails | The location of tails and related items should be matched left & right, or placed at center. |

**Experiment:** Tape a weight (such as a quarter) to one side of a kite far from the center. Does the size of the kite make a difference? Does the amount of weight make a difference?

**Experiment:** Cut a small hole in one side of the kite sail. Enlarge the hole and notice the effect on flight.

**Experiment:** Use a smaller diameter for the spreader on one side of the kite. How does the kite react?

**Q:** I'd like to attach lights to my kite. Where should I place the lights and the battery pack?

**A:** An easy solution is to add them to the line instead of the bridle or kite. If you do add them to the kite, make sure they are symmetric by placing them on the left, right, top, and bottom. Keep them light. It would be wise to place the heaviest part, the battery pack, at the center.

## Check the balance

Kites teach you about many aspects of balance. Symmetry is one form of balance. Flexibility is another. Weight is a third.

One way to balance a kite is to suspend it kite from a string. It should hang flat.

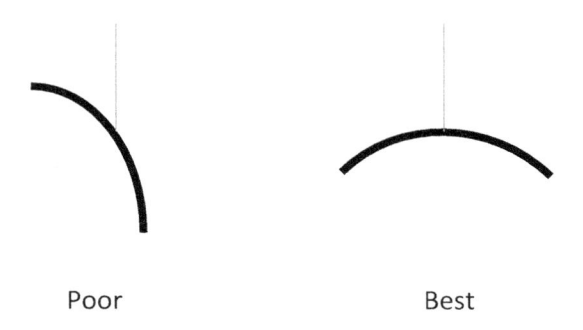

Poor                    Best

How can I improve upon the poor condition shown on the left?

- If your kite has a sliding bridle, adjust it.
- Move the bridle location to the correct spot.
- Balance by adding weight (such as tape) to the outer edge of the kite. That's usually the wing tip.
- If one side is longer than the other then shorten the long side. The sides should be identical in length.

Your kite should also be balanced front to back. For details, see the section called: Gliding vs. stalling.

# Case studies and their solutions

**Case Study:** A kite was spinning and crashing.

Problem: The kite needed the proper angle of attack and additional drag. This is a very common problem.

Solution: Adjust the bridle up or down in ¼" steps. Then add additional tails or longer tails.


**Case Study:** A festival workshop was building folded paper kites that wouldn't fly.

Problem: The spreader was too stiff, and there was no dihedral or bow.

Solution: Used dihedral and a flexible spreader to spill the wind.


**Case Study:** A festival provided printed sheets to build a sled kite that wouldn't fly.

Problem: The printer tried to fit everything onto the same sheet and changed the shape of the sled. They did not afterward.

Solution: We corrected their proportions and the sheets needed to be reprinted.


**Case Study:** Kites weren't flying steadily as wind increased.

Problem: Short tails were provided that had been cut to size. They were too short for the current wind speed.

Solution: We added a second tail to each kite to handle the stronger wind conditions.

**Case Study:** An Eddy kite wouldn't fly.

Problem: The consumer did not follow the instructions to prepare the kite for flight. Many people make the mistake of attaching the flying line to the back!

Solution: We removed the rubber band that was put around the packaged tails. We found the missing spars and inserted them correctly and attached the flying line to the front.


**Case Study:** A brand new airfoil kite included a drogue. In flight, the kite was swaying.

Problem: The aspect ratio was too high. That means the kite was too wide to be stable.

Solution: We lengthened the line to the drogue moving its drag further below the kite.


**Case Study:** A teacher was building a miniature kite with kids but it was unable to fly.

Problem: The sail loading wasn't right. For the size of the kite the paper was too heavy.

Solution: We recommended they switch to a lightweight tissue paper instead of copy paper.


**Case study:** A fox kite was not getting lift.

Problem: The wind was strong enough to push back both sides of the kite turning it into a narrow sliver and causing it to fall.

Solution: Add a thin spreader at the top to stiffen the kite. After that, the kite was flown to the end of the line.

**Kite repair checklist**

Some kites fly best in light wind while others fly best in moderate wind speeds. High wind speeds tend to exaggerate problems with kites. If you need to adjust your kite for the current wind speed or repair your kite, here are important things to check:

- ☐ The assembly directions have been checked carefully.
- ☐ The assembled kite is symmetrical.
- ☐ All spars are inserted properly.
- ☐ Cracked or broken spars have been replaced with identical spars on both sides.
- ☐ The bridle is not tangled, knotted, or on the wrong side of the kite.
- ☐ The bridle sets the correct angle of attack for the kite.
- ☐ Attachments to the line do not change the angle of attack of the kite.
- ☐ The sail is not torn.
- ☐ All pockets are intact and they are not twisted.
- ☐ If tails are included, they are fully extended.
- ☐ Tails are attached at the bottom of the center of the kite, not where they change the angle of flight.
- ☐ For Delta kites, the spars should be slid all the way down to the wingtips.

## Actions taken in the field

Once the kite has been built and tested Kite Fliers usually do not have the time or the resources to make drastic changes to their kites in the field or at the beach.

Kite Fliers assume the kite is symmetric in many ways. We also assume the kite is balanced. If it does not fly, the symmetry and balance should be checked.

Some changes are quick and easy. Here are the actions that kite fliers take in practice. In order to match the wind they will:

| Switch kites | Switch to a kite that is better tuned for the current wind speed. |
|---|---|
| Fly at a higher altitude | The wind at a higher altitude may be stronger, steadier, and more consistent but you'll need more line. |
| Bow the kite sail more | Adding bow to the sail is a minor adjustment and some kites are designed to make this adjustment easy. |
| Add or remove tails and drogues | • Add them in strong wind<br>• Remove them in light wind |
| Change the angle of attack | • Moving the bridle point higher is less stable but more lift.<br>• Moving the bridle point lower is more stable but less lift.<br><br>Note: When you change the angle of attack you may need to add additional tails to the kite. |

# Uses and applications of kites

## Measuring the atmosphere

Kites can lift instruments to thousands of feet above the earth and remain in that position for hours. The instruments are attached to the kite line rather than to the kite. This allows the kite to fly at the proper angle.

Kites have been used to make atmospheric measurements for more than 100 years.

> **According to the Blue Hills Observatory:**
>
> "The kite instruments measured temperature, altitude using barometric pressure, humidity, and sometimes wind speed. The kite location compared to its tethering spot gave us wind direction."

Today, radiosondes are used to take measurements, and are regularly flown on weather balloons worldwide. They are small packages of instruments that can transmit temperature, pressure, humidity, and GPS location every second.

**Experiment:** See if you can find a small *hygrostat* or a *radiosonde*. Use it to measure the water content of clouds and fog or use it to chart humidity at different altitudes.

## Parachute releases

You can release candy attached to a parachute and time its descent. It's even more fun to release an entire bag of parachuting candy.

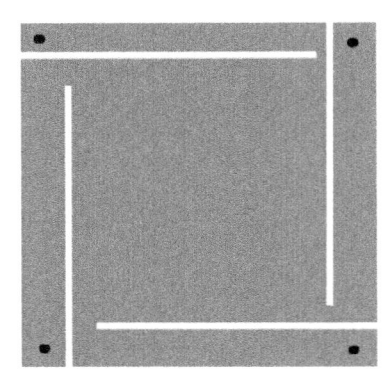

Trash bag parachute plan. Cut along the lines.

Mixing parachutes with helicopters and streamers allows candy to fall at different rates and spread out further downwind.

1. Fill a bag with each candy taped to a parachute, helicopter, or streamer.
2. Attach the bottom of the bag to the kite line with a carabiner.
3. Release by pulling a string that removes a pin from the top of the bag or use a radio controlled servo to pull the pin.

**Experiment:** In an area clear of people, release a bag of wrapped candy with parachutes, spinners, and streamers. Include some that are unaided. Time the descent of the first and last candy.

## Bear droppings

Increasing the lift on your kite allows you to have bigger payloads and drop large stuffed bears with a parachute!

Payloads can be dropped by pulling a cord or by radio control.

Here's an example of a bear wearing a backpack with a parachute:

The whole assembly hangs from a down rigger that allows it to be released from the kite line. Once the pin on the backpack has been pulled the bear falls gently to the ground.

It doesn't have to be a stuffed animal. You could release a glider or radio controlled helicopter from a high altitude.

**Experiment:** Drop an egg with a parachute. It must land safely.

## Kite Aerial Photography

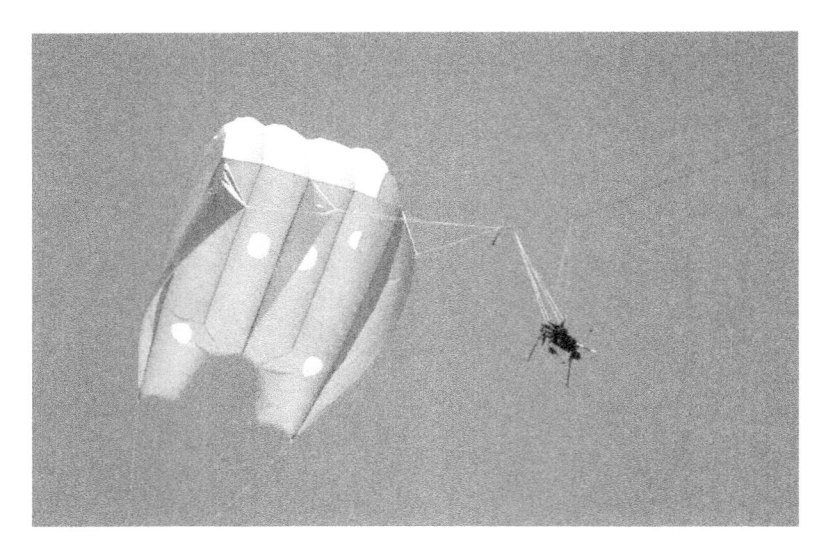

Using a kite to take photos is a popular technique called Kite Aerial Photography or "KAP" for short. A small camera can be mounted on the line that is attached to the kite.

By attaching the camera to the flying line instead of the kite it allows the kite to fly at the proper angle.

The camera is hung on a set of strings called a "picavet" that automatically keep the camera platform level. The angle of the camera can be adjusted with RC servos.

Letting out more line allows the camera to be lifted higher and retrieving the line allows the camera to take close-up photos and videos from above.

KAP has many applications. It is a low cost and silent way to record bird migration, do mapping, do environmental monitoring, estimate size, estimate distance, estimate area, or take photos of rooftop damage from hail.

**Experiment:** Use a kite to lift a cell phone while it records video.

## Wind instruments

Wind instruments include flutes that make a sound when the wind blows across them, a wind harp with strings, a hummer with strips that vibrate in the wind, or a wheel that spins and strikes a drum.

Example of a Taiwanese flying musical drummer.

Did you know that kites are occasionally flown at night with lights? With lights and wind instruments on your kite someone may conclude there's a UFO in the sky!

**Experiment:** Build a wind instrument that makes noise when the wind blows against it.

**Experiment:** Attach your instrument to a kite line and let it fly. Make sure the device will face the wind in flight.

## Line laundry

A popular practice is to attach decorations or elaborate windsocks to the line leading up to the kite. These come in a wide variety of shapes such as cats, dogs, frogs, dragons, pigs, owls, bears, and many cartoon characters. They can vary in size up to 100 feet (30 meters)!

**Note:** Adding decorations to the kite line may change the flying angle and cause your kite to overfly or dive when there is too much drag in the middle of the line. Your kite will need to be large enough to create enough lift to carry the additional weight and drag.

**Experiment:** Add a decoration to your kite line such as a flag. Make changes until you get the flag to fly horizontally.

## Line climbers

Line climbers are pushed by the wind and slide up the flying line to the kite. This is also called a "kite ferry."

Line climbers vary from simple pieces of paper that straddle the flying line to large umbrellas that spin as they ascend then close at the top only to speed down the same line they rode up to the kite.

Different mechanisms are used to gain lift, release, close, and return down the line.

Wind pushes this up to your kite line where it stops, closes, and returns.

**Experiment:** Make a simple line climber from a piece of paper. Attach it to your flying line and use a stopwatch to record the time it takes to reach your kite. Race against a friend!

## Ham Radio Antenna

There is a long history of using a kite to lift a radio antenna. The practice goes back to the beginning of radio.

In WWII naval fliers carried an emergency radio called a Gibson Girl. The antenna was carried aloft by a box kite. A 1/2 or 1/4 wave antenna at 500 kHz is a long piece of wire, so it was lofted from a life raft by a box kite. The waterproof kit came with a crank transmitter so the pilot could request a rescue.

During the Newport Kite Festival in 2017, Steve Uckerman used an airfoil kite to lift an inverted V, 5/8 wave antenna fed with a balanced ladder line to the height of 80'. Using that antenna, a solar panel for power, and a modern Ham radio set he was able to reach Italy, the U.K., Czech Republic, Germany, Belgium, and California.

## Kites and balloons

Both kites and balloons generate lift but they do so in vastly different ways. A balloon is usually filled with hot air or helium which is lighter than the surrounding air. That causes the balloon to rise. A kite gets lift from the force of the wind.

**Q:** Could we combine kites and balloons?

**A:** Yes. According to Wikipedia: "A kite balloon is a tethered balloon which is aerodynamically stabilized in windy conditions using similar principles to a kite. It typically comprises a streamlined envelope with stabilizing features and a harness or yoke connecting it to the main tether.

Kite balloons are able to fly in higher winds than ordinary round balloons and were extensively used for naval and military observation during World War I.

**Q:** Could we use balloons to lift kites to an altitude where they can fly?

**A:** Yes, that's been done. It looks funny, but on a day without wind it's a common thought.

One combination of kite and balloon is called a "Helikite." You can find out more here: http://www.allsopp.co.uk/

**Q:** Would it be possible to inflate sections on a kite?

**A:** Yes, this is how kitesurfing kites are built. The leading edge is inflated with air so it does not sink. Those kites can be launched and relaunched from water.

**Experiment:** Research a "Type M Kite Balloon."

## Summary of kite physics

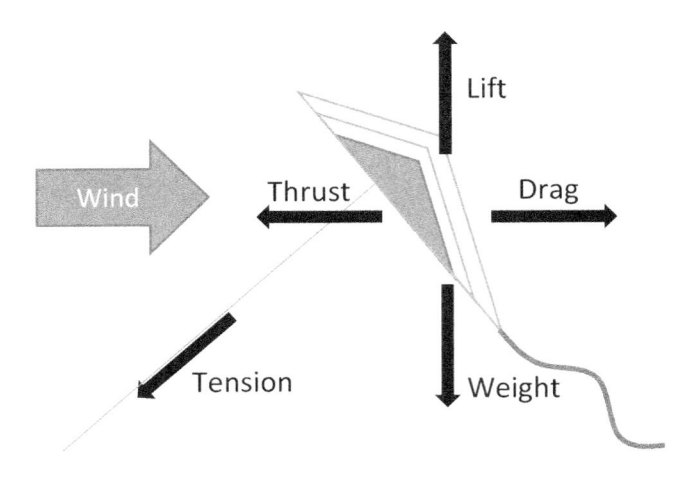

In the illustration above we see:

- The wind lifts the kite because of the angle of attack
- The kite can be different sizes and weights
- Lift opposes weight
- Drag opposes thrust
- The lift and drag form the tension on the line

Summarizing the methods of increasing stability include:

- Adding tails and drogues to increase drag for increased wind stability
- A kite that flexes allows heavy wind stability
- Dihedral or bow can be added
- Cells allow wind to pass through
- Vents in a kite reduce pressure on the sail and increase stability
- Properly positioning tow point and angle of attack

- Using an aerodynamic <u>shape</u>
- Symmetry and balance are both important
- For additional details, see the section called, "Methods of increasing kite stability."

**Experiment:** Make changes to the forces and motion of a kite that result in observable results but still allow the kite to fly. One of the objectives in 8th Grade science to *"Investigate the principles used to engineer changes in forces and motion."*

---

**Did you know?**

The Wright brothers tested their theories of flight using kites from 1900 to 1902.

# Glossary of terms

airfoil kite – typically a soft kite, this kite uses an airfoil shape to generate lift

angle of attack – the angle of the kite into the wind

aspect ratio – the ratio of the height to the width of the kite

bow – the curve of the kite frequently created with a tension line

bridle – the lines that connect to the kite to support the sail

bridle point – the point where the bridles come together

dihedral angle – the "V" shape of the kite

drogue – a bucket-shape that adds drag instead of tails

flying line – the line that connects the kite to the ground

frame – the spars that collectively give the kite shape

pitch – rotation around the horizontal-axis of a kite

roll – rotation around the longitudinal-axis of a kite

soft kite – a kite without spars

spar – the sticks that form the frame of the kite

Spectra® – a low-stretch material used for kite line

tension – the pull of the kite

thrust – pulling the kite into the wind

turbulence – rough air typically found near or around an object

yaw – rotation around the vertical-axis of a kite

# Kite physics resources

Books and educational kite resources:
www.kitingusa.com

Why does a kite fly?
http://www.gombergkites.com/nkm/why.html

The forces on a kite:
https://www.grc.nasa.gov/WWW/K-12/airplane/kitefor.html

The Drachen Foundation:
http://www.drachen.org/

Newton's First Law applied to kites:
https://www.grc.nasa.gov/WWW/K-12/airplane/newton1k.html

NASA interactive kite modeler:
https://www.grc.nasa.gov/WWW/K-12/airplane/kiteprog.html

Kite design basics:
http://wiki.dtonline.org/index.php/Kite_Design_Basics

NASA guide to kites:
https://www.grc.nasa.gov/WWW/k-12/airplane/bgk.html

Airfoil misconception explained:
http://amasci.com/wing/airfoil.html

# Bibliography

*The Tao of Kite Flying: The Dynamics of Tethered Flight* by Harm Van Veen, 1996

*Kites: The Science and the Wonder* by Toshio Ito and Hirotsugu Komura, 1983

*Building Free and Recycled Kites* by Glenn Davison, 2017

*Kite Workshop Handbook* by Glenn Davison, July, 2015

*Miniature Kites* by Glenn Davison, January, 2015

"A ship in port is safe; but that is not what ships are built for. Sail out to sea and do new things."
-Grace Hopper

Printed in Great Britain
by Amazon